EMERGING TRENDS IN PSYCHOLOGY, LAW, COMMUNICATION
STUDIES, CULTURE, RELIGION, AND LITERATURE IN THE GLOBAL
DIGITAL REVOLUTION

PROCEEDINGS OF THE 1ST INTERNATIONAL CONFERENCE ON SOCIAL SCIENCES SERIES: PSYCHOLOGY, LAW, COMMUNICATION STUDIES, CULTURE, RELIGION, AND LITERATURE (SOSCIS 2019), SEMARANG, INDONESIA, 10 JULY 2019

Emerging Trends in Psychology, Law, Communication Studies, Culture, Religion, and Literature in the Global Digital Revolution

Edited by

Yulianto Budi Setiawan
University of Semarang, Indonesia

Santi Rahmawati
Research Synergy Foundation, Bandung, Indonesia

Taylor & Francis Group

LONDON AND NEW YORK

Routledge is an imprint of the Taylor & Francis Group, an informa business

© 2020 Taylor & Francis Group, London, UK

Typeset by Integra Software Services Pvt. Ltd., Pondicherry, India

Publisher's Note
The publisher has gone to great lengths to ensure the quality of this reprint but points out that some imperfections in the original copies may be apparent.

Library of Congress Cataloging-in-Publication Data
Names: International Conference on Social Sciences Series: Psychology, Law, Communication Studies, Culture, Religion and Literature (1st : 2019 : Semarang, Indonesia) | Setiawan, Yulianto Budi, editor. | Rahmawati, Santi, editor.
Title: Emerging trends in psychology, law, communication studies, culture, religion, and literature in the global digital revolution : proceedings of the 1st International Conference on Social Sciences Series: Psychology, Law, Communication Studies, Culture, Religion and Literature (SOSCIS 2019), Semarang, Indonesia, 10 July 2019/ editors, Yulianto Budi Setiawan, University of Semarang, Indonesia, Santi Rahmawati, Research Synergy Foundation, Bandung, Indonesia.
Other titles: Proceedings of the 1st International Conference on Social Sciences Series: Psychology, Law, Communication Studies, Culture, Religion and Literature (SOSCIS 2019), Semarang, Indonesia, 10 July 2019
Description: London ; New York : Routledge, [2020] | Includes bibliographical references and index.
Identifiers: LCCN 2020005868 (print) | LCCN 2020005869 (ebook) | ISBN 9780367338367 (hardback) | ISBN 9780429322259 (ebook)
Subjects: LCSH: Social sciences--Research--Congresses.
Classification: LCC H62.A3 I575 2019 (print) | LCC H62.A3 (ebook) | DDC 303.48/33--dc23
LC record available at https://lccn.loc.gov/2020005868
LC ebook record available at https://lccn.loc.gov/2020005869

Published by: CRC Press/Balkema
           Schipholweg 107C, 2316XC Leiden, The Netherlands
           e-mail: Pub.NL@taylorandfrancis.com
           www.crcpress.com – www.taylorandfrancis.com

First issued in paperback 2021

ISBN 13: 978-1-03-224216-3 (pbk)
ISBN 13: 978-0-367-33836-7 (hbk)

DOI: https://doi.org/10.1201/9780429322259

Emerging Trends in Psychology, Law, Communication Studies, Culture, Religion, and Literature in the Global Digital Revolution – Setiawan & Rahmawati (eds)
© 2020 Taylor & Francis Group, London, ISBN 978-1-03-224216-3

Table of contents

Foreword   vii

Organizing Committee   ix

Scientific Committee   xi

Analysis of personality and behavioral performance   1
H. Widhiastuti

The role of leaders in cultivating Islamic values in companies   5
S.N. Hakim & M. Hanifa

The effects of positive and negative emotion on viral marketing effectiveness   9
D. Satrio, S.H. Priyanto & A.K.N.A. Nugraha

Teachers' revelations on self-motivation   14
A.C. Bocar, A.W. Rachmawati & S. Rahmawati

Legal protection for consumers in online transactions   18
D. Triasih, B.R. Heryanti & E. Pujiastuti

Legal protection of passing-off well-known marks for the welfare of consumers in Indonesia 22
D. Kridasaksana

Resilience in adolescents infected with HIV/AIDS   26
A.D. Savitri & Purwaningtyastuti

Reconstruction of the legal protection concept Janda Cerai
in the Batak Toba community based on the value of justice   29
A.P. Sihotang

Duality of structure in the Family Welfare and Empowerment organization (PKK)   36
I. Wiendijarti, H.I. Wahyuni & R. Witjaksono

Blonde hair as symbol of Patrick Bateman's displacement
in *American Psycho*: Semiotic approach   41
D. Anggraheni

Authority of religious court in settlement of the Shariah banking dissolution   45
D.I. Astanti, B.R. Heryanti & S.R. Juita

Implementation of well-designed interior spaces, with psychological concepts and solutions
for private workspaces in Surabaya   49
S.M. Sari & S. de Yong

Small industry and the relationship with protecting and managing the environment   56
D.T. Muryati, E. Pujiastuti & T. Mulyani

Redefining the relationship of majority and minority as a social principle   60
J. Basyir, R.F. Marta & Y.B. Setiawan

The scooter motorbike market monopoly by Honda and Yamaha in Indonesia, 2016–2019   64
Rr.C.W. Putri

Supervision of environmental permits as a juridical instrument for protection environment
based on community   68
E. Pujiastuti, D.T. Muryati & T. Mulyani

Malacca Portuguese intangible cultural heritage: An approach to cultural mapping   73
B.C. Rego, J.C. Bastos & D.N.A. Janie

Controlling the emotions of children with autism with social stories while at school   78
M.N. Irawan & T.K. Wai

Employability factor to improve readiness for changes   83
A.L. Kadiyono, R.A. Sulistiobudi & M.F. Abdurrohman

The challenges of the online journalism in the industrial revolution 4.0 era in Indonesia   90
A.P. Lestari, S. Solikhati & Y.B. Setiawan

Responding to the circulation of hoaxes using media literacy and information culture   95
T. Mulyani, D.T. Muryati & E. Pujiastuti

School resilience and religious radicalism in senior high schools   98
J.H.G. Purwasih & A.A. Widianto

Implementation of occupational health and safety management system in human resources
development to improve performance   102
H. Widhiastuti, G. Yuliasih & Y. Kurniawan

The influence of WhatsApp on improvements for fish farmers: A lesson from Semarang City,
Indonesia   107
F. Apresia, T. Elfitasari & T. Susilowati

The pragmatic force of expressive speech acts of the Banyumasan Javanese language in sale
and purchase transactions in Pasar Wage, Purwokerto Timur District, Banyumas Regency   111
M. Riyanton, M.T. Kariadi & E. Pujihastuti

Reconstruction of regulations dealing with the period of the presidents' term to minimize the
abuse of power   117
Sukimin

Sociocultural theory applied to collaborative learning in writing strategies   121
T.D. Wijayatiningsih & M. T. Kariadi

Author index   126

Emerging Trends in Psychology, Law, Communication Studies, Culture, Religion, and Literature in the Global Digital Revolution – Setiawan & Rahmawati (eds)
© 2020 Taylor & Francis Group, London, ISBN 978-1-03-224216-3

Foreword

The digital revolution is a change from analog mechanical and electronic technology to digital technology, which has happened since around 40 years ago. The impact of the development of digital technology has brought great changes to the civilization of world life.

Various parties feel facilitated in accessing information. All information regarding issues of Psychology, Law, Communication Studies, Culture, Religious, and Literature, can be easily accessed by anyone anywhere and anytime. However, with the development of technology, even more, negative impacts are felt rather than positive impacts. This can be seen from the existence of information that is excessive and not based on facts (hoaxes) in digital media so that it can provoke people to take wrong actions.

Through this international conference SOSCIS (Social Sciences Series: Psychology, Law, Communication Studies, Culture, Religious, and Literature), we raised the theme "Responding to Digital Revolution." This conference contains research articles that discuss the impact and solutions in facing the development of the digital revolution. We hope this conference can answer the needs of the community.

Regards
Yuliyanto Budi Setiawan
Santi Rahmawati

*Emerging Trends in Psychology, Law, Communication Studies, Culture, Religion,
and Literature in the Global Digital Revolution – Setiawan & Rahmawati (eds)*
© 2020 Taylor & Francis Group, London, ISBN 978-1-03-224216-3

Organizing Committee

General Chair
Dyah Nirmala Arum Janie

General Co-Chairs
Aria Hendrawan
Anna Dian Savitri
Subaidah Ratna Juita
Tatas Transinata
Hendrati Dwi Mulyaningsih

Conference Coordinator
Santi Rahmawati
Ani Rachmawati
Febrialdy Hendratawan

Conference Support
Almas Nabili Imanina

Information and Technology Support by Scholarvein Team

Emerging Trends in Psychology, Law, Communication Studies, Culture, Religion, and Literature in the Global Digital Revolution – Setiawan & Rahmawati (eds)
© 2020 Taylor & Francis Group, London, ISBN 978-1-03-224216-3

Scientific Committee

Editors:

Yuliyanto Budi Setiawan
Universitas Semarang

Anna Dian Savitri
Universitas Semarang

Subaidah Ratna Juita
Universitas Semarang

Scientific Reviewers:

Jaggapan Cadchumsang
Khon Kaen University, Thailand

Rukchanok Chumnanmak
Khon Kaen University, Thailand

Mochamad Chaerul Latif
Universitas Semarang

Piyanard Ungkawanichakul
Srinakharinwirot University, Thailand

Ali Ghufron
IKIP PGRI Bojonegoro, Indonesia

Putu Doddy Sutrisna
University of Surabaya, Indonesia

Rini Sugiarti
Universitas Semarang

Sulistyawati
Ahmad Dahlan University, Indonesia

Haslinda Anriani
Tadulako University Palu, Indonesia

Paripat Pairat
King Mongkut's University of Technology
North Bangkok, Thailand

J.Vignesh Kumar
National Institute for Research in Tuberculosis (NIRT), Indian Council of Medical Research (ICMR), India

Hardani Widhiastuti
Universitas Semarang

Jerny Dase
Hasanuddin University, Indonesia

Kim Alvin C. De Lara
Department of Education- Division of Rizal, Philippines

Amri Panahatan Sihotang
Universitas Semarang

Rohadi
Universitas Semarang

Bhargavi Kaveti
Kakatiya University, Telangana India

Liberty Nyete
University of Venda, South Africa

Tay Kok Wai
Universiti Kebangsaan Malaysia, Malaysia

Sruthi V.S.
Jawaharlal Nehru University, India

Garry Kuan Pei Ern
Universiti Sains Malaysia, Malaysia

Endah Pujiastuti
Universitas Semarang

Emerging Trends in Psychology, Law, Communication Studies, Culture, Religion, and Literature in the Global Digital Revolution – Setiawan & Rahmawati (eds)
© 2020 Taylor & Francis Group, London, ISBN 978-1-03-224216-3

Analysis of personality and behavioral performance

Hardani Widhiastuti
Universitas Semarang, Semarang, Indonesia

ABSTRACT: This article researched the influence of personality, as well as with the Big Five personality traits – extraversion, agreeableness, conscientiousness, neuroticism, and openness to experiences – on work performance in terms of behavior and results, in this case, at PT Jamsostek region of Central Java and DI Yogyakarta region using 186 subjects. The results showed a positive influence of extraversion and agreeableness on job performance and behavior. Neuroticism negatively affected results, and openness to experiences affected behavior positively. Measurement results have met the criteria for goodness of fit, among others, indicated by the value of the chi-square = 4.274; probability = 0.233; GFI = 0.992; AGFI = 0.878; TLI = 0.996; CFI = 0.955; RMSEA = 0.061; CMIN/Df = 1.425, besides the results of the regression coefficients that describe the influence on performance behavior on work of –0.17. Based on that account, the account officer must provide good service to customers at all times.

Keywords: personality, behavior performance, outcome

1 INTRODUCTION

The impact of the economic crisis that began in 1998 has been felt by most industries, including banking and finance. Business was no exception and the state of business has become a safe measure of the ongoing economic crisis. The side that felt the impact most was human resources management.

Performance orientation is associated with positive performance on the part of an employee (Sonnentag, 2000). If the distribution performance of the individual factors can be accurately determined, extensive information can be obtained regardless of the theoretical and practical contents generated in addition to variation and distribution of slope performance. Also alleged to be very closely related with the performance of particular behaviors is personality. The theory of personality has many facets related to an individual's behavior. Five factor theory of personality or the Big Five personality theory highlights not only intelligence but also the stability of emotions, experiences, confidence, and socialization of an individual. Therefore, with a background of problems encountered, we have attempted to analyze whether the elements of one's personality determine performance results or behavior.

2 LITERATURE REVIEW

2.1 *Performance of services and behavior*

Performance evaluations effectively support the implementation of strategic business, when employees understand what is evaluated, and the relevance of aspects that are rated, and they perceive the assessment as fair and valid (Robbins, 2002; Kreitner, 2005; Ivancevich, Konopaske and Matteson, 2008; Simamora, 2006).

According to Gatewood and Hubert (2001), a number of things can affect performance ratings at any given time, including (1) the consistency of performance; (2) criteria that consist of individual difference factors; (3) the consistency of the evaluation of the changes caused by the person's relevance to the organization, such as the purpose of the work and design work; and (4) reliability of measurement.

In this case, it really depends on the measurement techniques used at the time and place. In the opinion of Komaki et al. (2000) performance aspects used in his research under the title "A Rich and Rigorous Examination of Applied Behavior Analysis Research in The World of Work," are as follows: (1) production, (2) accuracy of the (3) security, (4) attention to the customer or client, (5) customer service, (6) results report, (7) supervision, (8) the result of the deposit, and (9) the seller. The research bases the analysis on an applied behavioral approach to learn the performance expectations, especially behavioral relationships and their consequences, consequent opportunities to take action, and consequent thoughtful thinking in relation to overall performance.

A major aspect of performance with service orientation according to the Dictionary University of Kentucky (2000) is: (1) Conformity and consistency with work and friendliness; (2) Effective use of interpersonal relationship skills; (3) Identifying service needs; and (4) willingness to respond.

Based on the main factors, a reflection of the behavior requires, among others: (1) Listening to customer problems and requests; (2) Making an extra effort in order to help the customer; (3) Anticipating customer needs and responding quickly, accurately, and agreeably; (4) Following up with customers and (5) empathizing with customers when problems occur.

Such factors are not separate from performance management, which must be implemented by the organization's management to affect both performance behavior and performance results. Performance management will have a positive impact for the organization, especially with respect to achievement of objectives of effectiveness and efficiency. As with Ruky's Opinion (2002) effective performance management programs should ensure, among other things: (1) relevance: things or factors measured are relevant (related) to the work, whether input, process, or output; (2) sensitivity: the system used should be sensitive enough to differentiate between employees who excel and those who do not excel; (3) reliability: the system used must be reliable, trusted using an objective benchmark, valid, accurate, consistent, and stable; (4) acceptance: the system used must be understandable and accepted by an employee who is administering an assessment or one who is assessed and facilitates the active and constructive communication between the two; (5) practicality: all instruments must be easy to use by both parties, not complicated or convoluted.

Greenberg and Baron (2000) mention, in addition to the five personality factors, another factor known to be important in relation to the success of performance, namely self-success, or an individual's belief about his skills to be demonstrated in success of specific tasks.

The three basic components of self-success include: (a) interests: a level of individual belief in what can be demonstrated; (b) strength: the usability of the performance demonstrated at a certain level; and (c) generalization: self-success concerns one situation or another task of the situation or another task.

2.2 *Five personality factors*

John (1990) abbreviated the five personality factors to OCEAN. With regards to the personality of the "top five", Digman (1990) has his own opinion that, among others, neuroticism encompassed negative feelings such as anxious, sad, easily touched, nervous. The openness factor encompasses the individual's complex openness, depth, and mental and life experiences. The Ekstraversi and the agreeableness factors include interpersonal cooperation and association with others. Lastly, is the so-called thoroughness factor, concerning task and achievement and control, which is social requirement.

Research conducted by Barrett (1996), Barrick and Mount (1991), Salgado (1997),, and 1990) proved that from five factors of personality only two factors are significantly related to performance, namely accuracy and extraversion. But this research is done at the manager level, so it needs to re-research the relationship between the five personality factors with performance. The research results of Yoo (2002) in Korea against all patterns of the accuracy, extraversion, and neuroticism factors, have a high correlation with the performance of results, compared with the agreeableness factor and the openness to experience factor. In the study, the result of the correlate of its coordinates approached 0 (ra =–0.01), and this is contrary to Hurtz and Donovan (2000) but highly consistent with European people results from Salgado (1997).

3   METHODOLOGY AND RESULTS

This research used the quantitative method. In the case of PT. Jamsostek, results have a relationship with the performance behavior of the sample of 186 account officers at PT. Jamsotek Central Java Indonesia. Logically, it can be said that if an account officer behaves well in the sense of appropriate SOTK and according to the norm, it will lead to increased performance results. This is evidenced by the research conducted in PT Jamsostek that if an account officer behaves well with his or her customers, then results based on earnings will also increase.

The data retrieval using the five-factor personality scale (Preliminary IPIP Scale) of Goldberg (1981) and Goldberg (1990) is a useful scale for measuring extraversion, agreeableness, thness, neuroticism, and openness to experience. The scale consists of 100 items divided into 20 items that reveal extraversion, 20 items that reveal the deal, 20 items that reveal thoroughness (conscienciousness), 20 items that reveal neuroticism, and 20 items revealing openness to experiences. Such calculations are noted that the CR value of all lines linking the influence of extraversion, agreeableness, and openness to the behavior and performance of results with CR above 1.96.

The structural equation model used to test the causality relationship between variables in the model shows a probability of 0.233 which indicates that the model is well received or, in other words, there is no difference between the matrix covariance. Measurement results have fulfilled the criteria of goodness of fit, among others, indicated by the value of chi-square = 4,274; Probability = 0,233; GFI = 0,992; AGFI = 0,878; TLI = 0,996; CFI = 0,955; RMSEA = 0,061; CMIN/Df = 1,425, Other than that, the result of a regression coefficient illustrates the effect of performance in a result of –0.17. Thus there is a negative influence between performances of behavior of account officer performance in the form of work result.

4   DISCUSSION

The results of the analysis showed that the variables extraversion and agreeableness affect the performance variables behavior and results variables conscienciousness and neuroticism affect the performance of results, and the variable openness to experience affects behavior performance.

The opposite result between theory and results of this research is likely because this research is done by analyzing the performance of behavior over three months. The assumption is that if a financial services account officer provides inadequate services to its customers, it will result in lowering the results achieved. The neuroticism variable has a negative effect on performance in the form of behavior. This is due to neuroticism facets such as anxiety, self-belittling, depression, quickness to take offense, sensitivity, and irritability which are not positive attributes in an account officer and do not support the work of an account officer, because indeed account officer jobs do not use many of the things that concern feelings.

Variable openness to experience has a positive influence on account officer behavior. Facets in the variable of openness to experience include, beauty, feelings, ideas, actions, and values, all of which can influence the account officer in responding to customer complaints as well as problem solving. Thus openness to experience has no influence with the performance of results, but is more influential in the performance of behavior because of the benefits of service felt directly by the customer.

Findings in the field suggest that account officers are sometimes faced with conflicting phenomena. On the one hand, account officers should maintain customer satisfaction by providing services as well as possible, which, as a consequence, is sometimes time-consuming. On the other hand, if the account officer only pursues the achievement of the target, the customers might be less satisfied.

Therefore, to face the competition of financial services companies, the account officer must still provide service to the customers as optimally as possible so that customers get satisfaction.

5 CONCLUSION

Based on the calculation process for data analysis and discussion, it can be concluded that personality affects the performance of account officers in the field of financial services. This is in accordance with the theories of Barrett (1996), Greenberg and Baron (2000), Salgado (1997), and Barrick and Mount (1991). Performance of the account officer in the form of behavior has a negative influence on performance in the form of work (outcomes). The findings in the field show that the account officers are sometimes faced with conflicting phenomena, on one hand attempting to maintain customer satisfaction, and on the other hand required to achieve targets that might lead to customers being less satisfied. Therefore, account officer selection should utilize personality tests, in order to obtained a personality picture of applicants and see if they accord to the personality of an account officer.

REFERENCES

Barrett, R. S. (1996) *Fair Employment Strategies in Human Resource Management*. Greenwood Publishing Group.

Barrick, M. R., and Mount, M. K. (1991) "The big five personality dimensions and job performance: A meta-analysis." *Personnel Psychology*. Wiley Online Library, 44(1), pp. 1–26.

Dictionary University of Kentucky (2000) *Services Job Performance, Dictionary University of Kentucky*. http://www.uky.edu/IR.

Digman, J. M. (1990) "Personality structure: Emergence of the five-factor model." *Annual Review of Psychology*. Annual Reviews Palo Alto, CA, 41(1), pp. 417–440.

Gatewood, R. D., and Hubert, S. (2001) *Field, Human Resource Selection*. South-Western.

Goldberg, L. R. (1981) *The Big Five Personality Dimensions*. http://www.rpi.edu/verwyc/BigFive oh.html.

Goldberg, L. R. (1990) "An alternative 'description of personality': The big-five factor structure." *Journal of Personality and Social Psychology*. American Psychological Association, 59(6), p. 1216.

Greenberg, J., and Baron, R. A. (2000) *Behavior in Organizations*. Prentice Hall.

Hurtz, G. M., and Donovan, J. J. (2000) "Personality and job performance: The Big Five revisited." *Journal of Applied Psychology*. American Psychological Association, 85(6), p. 869.

Ivancevich, J. M., Konopaske, R., and Matteson, M. T. (2008) "Perilaku dan Manajemen Organisasi. Jilid 1 dan 2." *Jakarta. Erlangga*.

John, O. P. (1990) "The Big Five factor taxonomy: Dimensions of personality in the natural language and in questionnaires." *Handbook of Personality, Theory and Research*.

Komaki, J. L. et al. (2000) "A rich and rigorous examination of applied behavior analysis research in the world of work." *International Review of Industrial and Organizational Psychology*, 15, pp. 265–368.

Kreitner, R. (2005) "Angelo Kinicki." *Perilaku Organisasi*, 2.

Robbins, S. P. (2002) "Prinsip-prinsip perilaku organisasi." *Jakarta: Erlangga*.

Ruky, A. S. (2002) *Sistem manajemen kinerja*. Gramedia Pustaka Utama.

Salgado, J. F. (1997) "The five factor model of personality and job performance in the European community." *Journal of Applied Psychology*. American Psychological Association, 82(1), p. 30.

Simamora, H. (2006) "Manajemen Sumber Daya Manusia." Edisi 2, STIE YKPN'. Yogyakarta.

Sonnentag, S. (2000) "Expertise at work: Experience and excellent performance." *International Review of Industrial and Organizational Psychology*, 15, pp. 223–264.

Emerging Trends in Psychology, Law, Communication Studies, Culture, Religion, and Literature in the Global Digital Revolution – Setiawan & Rahmawati (eds)
© 2020 Taylor & Francis Group, London, ISBN 978-1-03-224216-3

The role of leaders in cultivating Islamic values in companies

Siti Nurina Hakim & Milla Hanifa
Universitas Muhammadiyah Surakarta, Surakarta, Indonesia

ABSTRACT: Leaders make a significant contribution to the management of a company. The purpose of this research is to reveal the role of leaders in instilling Islamic values. A descriptive qualitative approach was employed. There were three main informants and seven supporting informants of this research. The data collection techniques were observation, in-depth interview and documentation. The results of this study reveal that the leaders have several main roles in instilling Islamic values. 1) "Model the way," by always giving advice and behaving exemplarily to the employees. 2) "Inspire a shared vision", where the leader ensures that employees understand the company vision. 3) "Challenge the process" within the organization, where leader performs monitor and build relationships with outside parties to ensure the development of human resources or employees. 4) "Enable others to act," by empowering them to take action and establish good relationships with employees. 5) "Encourage the hearts" by sincerely encouraging and motivating employees. Furthermore, the cultivation of Islamic values is done through the performance of Islamic rituals, including the five obligatory prayers in congregation, prayer time for customers, being patient and kind, celebrating the prophet's birthday, donating to orphanages, allocating 2.5% of income for charity, and with prohibition from drugs, alcohol, gambling and prostitution.

Keywords: Leader, Leaders Role, Cultivating Islamic Values

1   INTRODUCTION

The success or failure of a company is often linked to the role of its leaders. The main role of a leader is to influence others to deliberately and willingly achieve the goals of the company. Some leaders have attempted to cultivate Islamic values within their companies, yet not all are able to apply Islamic values based on the authentic Islamic teachings.

Generally, the implementation of Islamic values includes the obligation for employees to perform obligatory prayers in congregation and have religious lectures on a daily basis. Furthermore, the images of leader are reflected in the four primary roles of effective leadership (Nanus, 2001; Komariah, 2003; Sujatno, 2008; Daswati, 2012):

a. The role of determining the direction: a leader must be able to select and set specific targets by taking into account the external environment of the future, which is the empowerment of all organizational assets and resources to achieve the vision.
b. The role of agent of change: leaders provide inspiration for all transformation and development in the global environment and predict the implications for the organization, so as to scale priorities for change implied by the vision of organization.
c. The role of spokesperson: leaders must be able to negotiate and create external networks, develop and communicate a vision, and empower and make changes.
d. The role of trainer: leaders must be able to communicate the current reality to others, including the vision and the measures to realize it. A leader should encourage others to move forward and guide them to actualize the potential based on the vision.

Kouzer and Posner in Kaswan (2013) suggested aspects attached to the role of leader: model the way, inspire a shared vision, challenge the process, enable others to act, and encourage the hearts. Furthermore, Hakim (2012) explained that the aspects of Islamic values are essentially distinguished as follows:

a. The values of faith that teach humans to believe in Allah.
b. The values of worship that teach human beings to always be sincere in their actions.
c. The character values that provide guidelines for good attitude and behaviour.

PO Haryanto is an autobus company that has implemented several regulations based on Islamic character values. The regulations include the obligation to perform five daily prayers in congregation, the allocation of 2.5% of income for charity, donations to orphanags, and the prohibition of drugs, alcohol, gambling, and prostitution.

2 METHOD

The selection of the main informants used a purposive sampling technique with several criteria: Muslim, leader with at least two employees, and a working period of at least three years. Moreover, seven supporting informants were also included in this study, who were the subordinates of the selected leaders. Data collection techniques were non-participant observation, in-depth interview and documentation. Interviews were conducted to disclose the role of leaders in instilling Islamic values in the company.

3 RESULTS AND DISCUSSION

3.1 *The role of leader in cultivating Islamic values in PO Haryanto*

The role of the leader can be distinguished in five aspects as suggested by Kouzer and Posner in Kaswan (2013). A leader has the role of inspiring a shared vision, as demonstrated by providing good examples to employees. The leaders in PO Haryanto demonstrated this by implementing the applicable regulation of practicing prayers in congregation. Najib (2015) affirmed that a leader is expected to be a good influence and guide others to achieve goals. According to Tafsir Ibn Kathir of Surah An-Nahl verses 120–121, the idolaters or leaders are those who guide to goodness and carry out all the commands of Allah as reflected in worshiping Allah alone.However, not all employees respond the regulation of performing five daily prayers in congregation. Some were not encouraged to perform prayers in congregation or take a break for prayer. They might be unfamiliar with such a habit, particularly the drivers, driver assistants and mechanics and it was reported that some resigned due to this regulation. In Tafsir Al-Misbah of Surah An-Nisa verses 83, ulil amri has the authority in managing the affairs of others as long as they are believers and the regulation does not conflict with the command of Allah and the Prophet (Jamal, 2014).

3.2 *The second aspect is to inspire a shared vision*

The leader must communicate the company vision to the employee during the signing of the employee agreement. In addition, the leader should provide advice, behave exemplarily and emphasis to ensure employees carry out the company vision accurately. A leader may be the first person to convey the company vision and Richard L. Draft emphasized this aspect in which a leader has legitimate power for using their authority (Fahmi, 2012).

Nasrudin (2010) explained the criteria that should be possessed by a leader, including the communication skills to convey ideas concisely, clearly and appropriately. Nanus (2001), Komariah (2003) and Sujatno (2008) proposed that leaders have a role as trainers or tutors, and should be able to communicate a vision and how to realize it in others, as well as regularly encourage others to progress and guide on how to actualize the potential to achieve a vision (Daswati, 2012).

3.3 *The third aspect is challenging the process*

A leader must provide supervision and monitoring of employees' performance, both directly and indirectly. Najib (2015) proposed that leaders should have the skills to influence and provide direction in order to achieve goals. The leaders of PO Haryanto have challenged themselves to initiate a network with external parties, which can be a way to ensure employees learn and develop their potentials. Mintzberg suggested the role of leaders as an intermediary or mediator, in which they have to integrate with staff, colleagues and external parties in order to obtain information, so that they serve as the receiver and collector of information and understand all aspects related to their company (Thoha, 1996).

3.4 *The fourth aspect is enabling others to act*

The leaders of PO Haryanto treat all employees as a big family. This leads to the absence of discrepancies or discrimination among employees. Consequently, employees are not reluctant to raise any objections and leaders easily understand them and prioritize their loyalty. This contradicts Kouzer and Posner, that leaders must develop cooperation and collaboration, appreciate and respect employees, and entrusts them to perform duties (Kaswan, 2013). Leaders are expected to monitor their employees on a regular basis and conduct briefings and meetings if there is a problem. Rivai and Bacthiar (2013) suggested that leadership is the head, chairman, manager or director of an organization, and someone who possesses skills and excellence in their field, and ultimately is able to influence others to collectively carry out certain activities to achieve several goals.

3.5 *The fifth aspect is encouraging the hearts*

The leaders of PO Haryanto always motivate their employees and reward them. Mintzberg suggested that leaders shall conduct interpersonal relations with employees by performing their primary functions, namely leading and motivating employees as well as managing the organization (Thoha, 1996). Kouzer and Posner explained the role of the leader in encouraging their hearts. Leaders always bring hope, contentment, enthusiasm and motivation (Kaswan, 2013). This is also in accordance with Tafsir Ibn Kathir of Surah An-Nahl: 120–121, that al-ummah means idolator or leader who teaches goodness and carries out all commands of Allah (Rivai and Bacthiar, 2013).

3.6 *Islamic values promoted in PO Haryanto*

Several Islamic values have been promoted in PO Haryanto: obligation to perform five daily prayers in congregation, the habit of being patient and generous, religious lectures to celebrate the prophet's birth, donations to orphanages, the allocation of 2.5% of income for charity, and the prohibition of drugs, alcohol, gambling and prostitution. The cultivation of Islamic values in PO Haryanto has implications for company achievements. This has been developed as a company, accompanied with the improvement of morality and spiritual contentment obtained by employees.

Halstead (2007) asserted the basic values of Islam, namely morality characteristics that refer to the duties and responsibility regarding Islamic law and Islamic education in general. Hakim (2012) verified the aspects of Islamic religious values, characteristics or moral values, by providing and directing people to behave based on their rights and good character. Similarly, Kurnialoh (2015) emphasized the spiritual dimensions of Islamic values, namely faith, piety, and noble character as reflected in worship and muamalah.

On the contrary, employees who refuse to perform according to company regulations have actually denied the command of Allah revealed in Surah An-Nisa: 59 as elaborated in the Tafsir Kalam al-Manna, that Allah obliges humans to obey their leader, since it would not be perfect one's world affairs and religion except one's submission and obedience to the leader

(Jamal, 2014). Richard L. Daft reinforced the claim that leaders have coercive power based on the authority to punish or recommend a penalty (Fahmi, 2012; Hikmat, 2009).

4 CONCLUSION

A study of the role of leaders in cultivating Islamic values in the company was carried out. The results reveal that the role of leaders can be distinguished into five practices: model a way; inspire a shared vision; challenge the process by initiating networks with external parties; enable others to act by working under family principles; and encourage the hearts. The Islamic values promoted and implemented in PO Haryanto are emphasized by character or moral values, namely the obligation to perform five daily prayers (in congregation), the allocation of 2.5% of income for charity, regular donations to orphanage, and the prohibition of alcohol, gambling, drugs and prostitution for employees.

REFERENCES

Daswati, D. (2012) Implementasi Peran Kepemimpinan dengan Gaya Kepemimpinan Menuju Kesuksesan Organisasi, *Academica*, 4, 1.

Fahmi, I. (2012) Manajemen Kepemimpinan Teori dan Aplikasi, *Bandung*: *Alfabeta*.

Hakim, L. (2012) Internalisasi nilai-nilai agama islam dalam pembentukan sikap dan perilaku siswa Sekolah Dasar Islam Terpadu Al-Muttaqin Kota Tasikmalaya, *Jurnal Pendidikan Agama Islam-Talim*, 10, 1, 67–77.

Halstead, J. M. (2007) Islamic values: a distinctive framework for moral education?, *Journal of Moral Education*. Taylor & Francis, 36, 3, 283–296.

Hikmat, H. (2009) *Manajemen pendidikan*. Pustaka Setia.

Jamal, K. (2014) Terminologi pemimpin dalam Al Quran (Studi analisis makna Ulil Amri dalam kajian tafsir tematik), *An-Nida*, 39, 1, 118–128.

Kaswan (2013) *Leadership and teamworking*. Bandung: Alfabeta.

Kurnialoh, N. (2015) Nilai-Nilai Pendidikan Agama Islam dalam Serat sastra Gendhing, *IBDA: Jurnal Kajian Islam dan Budaya*, 13, 1, 98–113.

Najib, M. (2015) Dinamika Kelompok, *Bandung*: *CV Pustaka Setia*.

Nasrudin, E. (2010) Psikologi manajemen, *Bandung*: *Pustaka Setia*.

Rivai, V. and Bacthiar, R. A. (2013) Pemimpin dan Kepemimpinan dalam Organisasi, *Jakarta*: *PT Raja Grafindo Persada*.

Thoha, H. M. C. (1996) *Kapita selekta pendidikan Islam*. Pustaka Pelajar.

Emerging Trends in Psychology, Law, Communication Studies, Culture, Religion,
and Literature in the Global Digital Revolution – Setiawan & Rahmawati (eds)
© 2020 Taylor & Francis Group, London, ISBN 978-1-03-224216-3

The effects of positive and negative emotion on viral marketing effectiveness

Danang Satrio, Sony Heru Priyanto & Albert K.N.A. Nugraha
Universitas Kristen Satya Wacana, Salatiga, Indonesia

ABSTRACT: Message content has become a critical factor for the success of viral marketing of cultural products. Appropriate content allows the receiver to remember the content, and this is particularly effective when emotion attributes are embedded in the content. A receiver might be willing to forward the viral content so it spreads in a way that may be 'endemic' and 'contagious'. The purpose of this study is to investigate the role of emotion attributes in the viral content that arouse emotion and, in turn, affect the intention to engage. The object of this study is viral marketing for Batik and this study makes use of experiment design. The result indicates that emotion attributes in viral content distinguish the intention to engage.

Keywords: viral Marketing, emotion, content, intention to engage

1 INTRODUCTION

Viral marketing has become a critical element of marketing strategy element (Koch and Benlian, 2015). Key performance indicators of viral marketing activities are measured by the uniqueness of visitors, the level of interaction, the relevant actions taken, the size of the intensity of the conversation, the credibility of the message giver, freshness, and relevance (Fisher, 2009). According to Scott (2011), appropriate content become a pivotal factor for the success of viral marketing. Viral content becomes effective when its distribution spreads quickly, such as contagious viral content (Petrescu and Korgaonkar, 2011). The nature of social media has become an arena for viral marketing activities, given its ability to facilitate information-sharing among individuals in the network (Miller and Lammas, 2010).

Emotion becomes one important variable in consumer behavior and is able to drive an individual to respond to environmental stimuli and act (Vainikka, 2015). An individual is more likely to engage in viral marketing when they are able to justify the benefit of the message or associate it with their values, interests, or beliefs. Such individuals may receive and share information such as news, articles, coupons (Berger and Milkman, 2012) in order to express their values, interests, beliefs, including emotions, to others. Empirical research shows that successful viral content must attach emotional attributes that motivate consumers to spread messages on their social networks (Dobele et al., 2007). The appropriate emotional attributes potentially attract consumers to see and appreciate the messages (Phelps et al., 2004). In the context of advertising, emotional tone in viral video influences intention to forward (Eckler and Bolls, 2011).

Previous studies indicate a lack of robustness of the emotion role to influence consumer responses (Berger and Milkman, 2012; Eckler and Bolls, 2011). Both positive and negative emotional attributes in the viral content might differ in how they arouse consumer emotional responses. Therefore, this study was designed to test the role emotional attributes in the message content has on consumer decisions, which is mediated by consumer emotional responses (i.e., positive and negative emotional attributes). Xavier and Summer (2009) show that previous studies have not described and explained the role of positive and negative emotions in the diffusion of viral content in the context of cultural product. The current study examines whether the type of emotion used in viral marketing campaigns for Batik as a cultural product distinguishes consumer behavior. Indonesia claims Batik as its heritage in 2009 and UNESCO recognized Batik as a Masterpiece of Oral and Intangible Heritage of Humanity (UNESCO, 2009).

2 LITERATURE REVIEW

2.1 Viral marketing

Viral marketing encourages people in their real life and social media networks to discuss companies and products that correspond with the definition from Helm (2000). Viral marketing has become an increasingly popular promotional tool for various brands (Eckler and Bolls, 2011). In reality, viral marketing is perceived to be cheaper (Woerdl et al., 2008) and can reach large numbers of message recipients relatively quickly (Cruz and Isi, 2008; Woerdl et al., 2008). In addition, viral messages are exempted from geographic and time boundaries, which allow their global distribution compared to conventional communications (Goldsmith and Horowitz, 2006).

2.2 Content

Content has become an important key success for viral marketing and, therefore, careful planning is key to crafting this content for social media. Moreover, Ward (2000) states that a messenger could only plan and disseminate the content and, unfortunately, the results are beyond their control (Sebastian, 2015). If the planning process is not executed properly, the diffusion process stops at the starting point. It is very important to articulate goals and objectives, marketing strategy and tactics into the planning of content creation.

2.3 Emotion

Emotions are strong and relatively uncontrollable feelings that affect individual behavior (Hawkins, 2010). The emotional connection between recipient and content plays an important role in increasing the likelihood of forwarding viral content (Dobele et al., 2007). However, the extent of its effect may differ depending on the type of emotional attributes. There are six primary emotions of surprise, joy, sadness, anger, fear and disgust, which can be further categorized into positive and negative emotions (Dobele et al., 2007).

2.4 Intention to engage

Intention variables capture the motivational elements that influence behavior (Perera & Dharmadasa, 2016). Ho and Dempsey (2009) state that the intention to disseminate viral content or not is made voluntarily. Their study indicates that internet users should pay attention to online content before they forward it. In this sense, intention expresses individual commitment to perform certain behaviors (Hsu & Lin, 2008) and relates it to attitudes and behavior preferences such as forwarding messages through social media (Okazaki, 2009).

Based on the relation between the content, emotion and intention to engage, the proposed viral marketing model is shown in Figure 1.

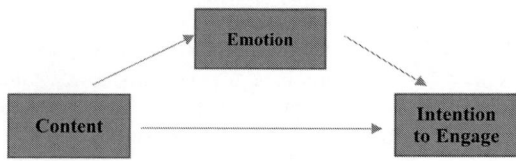

Figure 1.   The proposed viral marketing model.

3 RESEARCH METHODS

The current study applies experimental research design (Latipun, 2006), considered ideal due to its ability to clearly describe the influence of independent variables on dependent variables.

The research subjects were 120 students of the Faculty of Economics, Pekalongan University. The findings are supported by the knowledge of students as the "profession" that uses the internet the most compared to other sectors.

A set of interviews, discussions, and documentation from key informants was produced to collect necessary materials for content production. There are two content items that describe the messages about Batik, delivered in the form of short video clips. The contents differ based on the attachment of emotional attributes (i.e., positive and negative emotions). Subsequently, the experimental procedure was conducted and all experiment participants responded to the contents conveyed in the laboratory setting by filling out a self-administered questionnaire. The participant's responses were then evaluated through the use of an independent sample t test as a comparative test or a different test.

4 FINDINGS AND DISCUSSION

4.1 *Content*

The results indicate that both positive and negative content do not differ significantly as shown by the p value 0.249 (p>.05). Both content items resulted in the same score (13), which indicates that respondents could not distinguish the content based on emotional attributes. This implies that positive or negative emotional attributes in content might not influence other variables. This finding is in line with previous studies (Berger and Milkman, 2012; (Eckler and Bolls, 2011; Teixeira, 2012; Paus & Macchia, 2014; Berger and Milkman, 2012).

The plot of respondent' responses is shown in Figure 2.

4.2 *Emotion*

The following are the respondent' responses attributed to emotion variables. Figure 3 shows that emotional responses to both content items differ significantly, which is shown by a p-value of <.001. Emotional response to content with negative attributes had an average score of 25 while the response to content with positive attributes had an average score of 19, which is lower than the earlier response. This corresponds with the results from previous studies (Dobele *et al.*, 2007) although it is contradicts results from other studies (Chohan 2013; Eckler and Bolls, 2011; Libert and Tynski (2013) indicate that pleasant emotional tones have the strongest effect on forming attitudes and stimulating forwarding, while unpleasant emotional tones have the weakest effects.

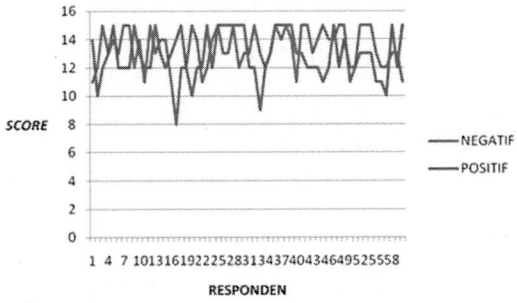

Figure 2. Respondents' responses to content.

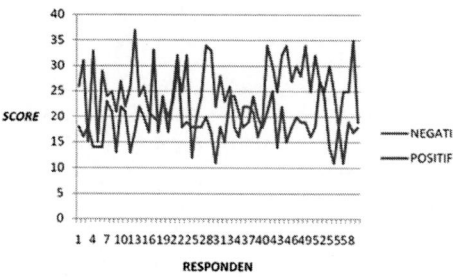

Figure 3.   Respondents' responses to emotion attributes.

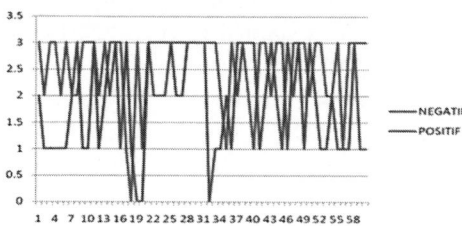

Figure 4.   Respondents' responses on intention to engage.

4.3   *Intention to engage variable*

Intention to engage in disseminating viral content becomes one important variable for marketers to measure the success of viral marketing campaigns. Interestingly, the result indicates that intention to engage differs significantly between content with positive and negative emotional attributes (p-value of .007). Intention to engage with content with negative attributes had an average score of 2.42, while engaging with content with positive attributes had an average score of 1.97. See Figure 4.

5   CONCLUSION, IMPLICATIONS AND RECOMMENDATIONS

Although participants did not feel a significant difference in the content variable (i.e., positive and negative attributes), emotional responses and intentions to engage were significantly different. This situation can occur because the level of emotion received can influence the intention to engage with others, and the higher the level of emotions involved the higher the desire is to share the content. Thus, it can be concluded that emotional relationships play an important role in effectively influencing viral message forwarding behavior. Marketers may consider including emotional attributes in viral content for cultural products such as Batik. This potentially increases the effect of viral content on the intention to engage and, in turn, increases the effectiveness of viral marketing campaigns. Further studies may elaborate on this by comparing the effect of emotional attributes among different product categories.

REFERENCES

Berger, J. and Milkman, K. L. 2012. What makes online content viral? *Journal of Marketing Research* 49, 2, 192–205.
Dobele, A. *et al.* 2007. Why pass on viral messages? Because they connect emotionally, *Business Horizons* 50, 4, 291–304.
Eckler, P. and Bolls, P. 2011. Spreading the virus: Emotional tone of viral advertising and its effect on forwarding intentions and attitudes', *Journal of Interactive Advertising*, 11, 2, 1–11.

Fisher, T. 2009. ROI in social media: A look at the arguments, *Journal of Database Marketing & Customer Strategy Management*, 16, 3, 189–195.

Helm, S. 2000. Viral marketing-establishing customer relationships by 'word-of-mouse', *Electronic markets*, 10, 3, 158–161.

Koch, O. F. and Benlian, A. 2015. Promotional tactics for online viral marketing campaigns: how scarcity and personalization affect seed stage referrals, *Journal of Interactive Marketing*, 32, 37–52.

Miller, R. and Lammas, N. 2010. Social media and its implications for viral marketing, *Asia Pacific Public Relations Journal*, 11, 1, 1–9.

Petrescu, M. and Korgaonkar, P. 2011. Viral advertising: Definitional review and synthesis, *Journal of Internet Commerce*. 10, 3, 208–226.

Phelps, J. E. *et al*. 2004. Viral marketing or electronic word-of-mouth advertising: Examining consumer responses and motivations to pass along email, *Journal of Advertising Research*, 44, 4, 333–348.

Scott, D. M. 2011. *The New Rules of Marketing & PR: How to Use Social Media*. Chichester: John Wiley & Sons.

Vainikka, B. 2015. *Psychological factors influencing consumer behaviour*. Centria ammattikorkeakoulu Keski-Pohjanmaan ammattikorkeakoulu.

Woerdl, M. *et al*. 2008. Internet-induced marketing techniques: Critical factors in viral marketing campaigns, *Journal of Business Science and Applied Management*, 3, 1,35–45.

Xavier, L. J. W. and Summer, G. Y. S. 2009. Viral marketing communication: the Internet word-of-mouth, a study on consumer perception and consumer response, Unpublished Master Thesis, Blekinge Institute of Technology.

Teachers' revelations on self-motivation

Anna C. Bocar
Gulf College, Muscat, Sultanate of Oman

Ani Wahyu Rachmawati
International Women University, Bandung, Indonesia Research Synergy Foundation, Indonesia

Santi Rahmawati
Research Synergy Foundation, Indonesia

ABSTRACT: The purpose of this study is to determine the level of teachers' self-motivation. The study employed quantitative-descriptive research design. The participants are male and female teachers from three countries, Philippines, Vietnam, and the Sultanate of Oman, including nationals and non-nationals. The instrument used in gathering the data was sourced from an online article and slight modifications were made to suit the participants. A four-point qualitative scale was used to measure the self-motivation level of the participants and the weighted mean was used for interpreting the results. The study found that males' self-motivation levels are higher than females' as propelled by 1) self-confidence and self-efficacy; both male and female participants showed high motivation levels driven by 2) positive thinking and positive thinking about the future as well as 3) focus and strong goals; and 4) environmental factors inspired both groups of participants infrequently. In conclusion, the overall results showed that these four factors influenced self-motivation levels most of the time, and thus the teachers are highly motivated. A high level of self-motivation is below the exceptional level of self-motivation, and hence this implies that more push factors are necessary to facilitate remarkable and unique output. Incomparable self-motivation attitudes could lead to the attainment of the goals.

1 INTRODUCTION

Justice Smith stated, "Stay focused and stay determined. Do not look to anyone else to be your determination – have self-determination. It will take you very far." This simply means that a person can reach what they are aiming for using self-rule. It promotes the idea that people can be productive and can go beyond the distance they can see. It also suggests self-motivation. Based on the dictionary definition, self-motivation refers to completing tasks continuously without supervision. The degree of activity connected with self-inspiration defines an individual's achieved goals, and their persuasive abilities, privileged circumstances, and capacities lead other people to expect they can accomplish these objectives. This could be true when one's desire is coupled with enough hard work to succeed (Goal Setting Tools, 2014).

2 LITERATURE REVIEW

C. Tagaylo affirmed that teaching has always been considered the noblest profession as it requires so much hard work and dedication. Only those who are highly motivated can dauntlessly take on the challenges and remain on the pedestal of the teaching profession. Thus, self-motivation is indeed significant in the teaching profession (personal communication, April 2, 2019). G. Jocson stated that, as a whole, attaining academic objectives can be attributed to a self-motivated educator. If a teacher is self-motivated then they will be happier, deliver their work effectively, and will be keen to learn and update their professional skills (personal communication, March 28, 2019). In the words of M. Tizon, teaching is not about how many learners a teacher has but how much the teacher leaves a spark in their lives. Thus, a teacher should have self-motivation that will emanate to and inspire the students (personal communication, April 2, 2019).

Motivation refers to the inner or mental process that helps a person complete activities to realize their objectives and goals. The most efficient and effective way of dealing with life's challenges is an appropriate mindset and self-motivation is a source of this. Thus, self-motivation plays a significant role in an individual's responsiveness. It is possible to learn great amounts when a person has intense self-motivation. Staying motivated is an essential element in pursuing learning, realizing objectives, achieving goals, and attaining success. Self-development could be a product of self-motivation as an extremely motivated person could have the ability to change their life (Siddiqui, 2016). I. Villareal expressed that self-motivation is very important in the teaching profession as it optimizes teaching productivity, opens avenues for creativity, and creates positivity that influences others to be more enthusiastic and more focused on teaching (personal communication, March 9, 2019). Fournier (2018) explained that some people believe they can influence events and outcomes, while others blame outside forces for everything. This concept is worth considering, however self-motivation is needed to make this thing happen. Julian Rotter in the 1950s, as cited in Neill (2006), stated that under the Locus of Control theory an individual perceives that the events in their life are caused by forces, from either inside (internal locus of control) or outside (external locus of control) oneself. This means that something happens because of some other thing that causes it to happen. Moreover, some perceptions suggest that internal locus of control is preferred because this would also mean that there is personal control and self-determination (Mamlin, Harris and Case, 2001; Hans, 2000). These two aspects are essential elements for self-motivation.

3   RESEARCH METHODS

The quantitative-descriptive design is utilized in this study. The participants are teachers from three countries, Philippines, Vietnam, and the Sultanate of Oman, including nationals and non-nationals. These participants comprise male and female teachers. The focus of this research is to determine the self-motivation level of the teachers, regardless of their location.

Moreover, after determining the level of participants' self-motivation level, the differences in these levels can also be identified. The instrument used for gathering the data was sourced from an online article (Goal Setting Tools, 2014), and slight modifications were made to suit the participants. To gather responses from the participants, the researchers utilized a four-point qualitative scale. A numeric value was assigned with its corresponding hypothetical mean range. For purposes of clear analysis, verbal interpretations were prepared to correspond to each qualitative scale of measurement. Furthermore, the weighted mean was employed to determine the final interpretation of the results.

3.1   *Data gathering procedure*

A survey questionnaire was posted online. A form in Google was created, and the indicators were entered along with the qualitative scale of measurement. The researcher shared the link with her colleagues in the Sultanate of Oman, and also sent it to advocate teachers in the Philippines and Vietnam, who forwarded it to their colleagues to complete.

4   RESULTS AND DISCUSSION

The 56 participants comprised 19 males and 37 females, and the majority were aged 36–50. Four main factors drive the level of participants' self-motivation. Each factor has indicators that measure their motivation level. The first factor of self-confidence and self-efficacy drives the male and female motivation at different levels: the male teachers revealed that they are highly motivated by these factors while the female teachers said they are only sometimes motivated by their self-confidence and self-efficacy. Male and female teachers' self-motivation levels are in the same range of motivated most of the time for the second factor of positive thinking and positive thinking about the future, and the third factor of focus and strong goals. The participants showed that they are optimistic thinkers, and have concentration and irrepressible attitudes when it comes to their goals. The participants revealed that the are usually highly motivated by these indicators. Also, both groups of participants demonstrated they are infrequently motivated by the fourth indicator, the working environment. Analyzing the strengths of these four factor indicators that contributed to the level of participants' self-motivation, the results show varying degree of impulse driven by each of the different factors. However, based on the final statistical measurement, both groups of participants are motivated most of the time. This signifies that their self-motivation is at a high level.

5   FINDINGS

Of the 56 participants, 37 were female and 19 were few males. The majority of the participants were aged 36–50 years. The males' self-motivation level as propelled by self-confidence and self-efficacy was higher (often) than the females (sometimes). The male and female participants showed high (often) self-motivation levels driven by positive thinking and positive thinking about the future as well as by focus and strong goals. The environmental factors inspired the self-motivation level of both groups of participants infrequently or only some of the time.

6   CONCLUSION

A person's talents, expertise, and capabilities are essential in the workplace. Moreover, their competencies and other technical expertise must be coupled with a sufficient degree of motivation to attain life's goals. This is particularly important for teachers, the profession that mentor the future leaders of society or the country in the wider scenario. In the teaching profession, motivation needs to be present for both the administration and teaching staff at academic institutions. Management is required to motivate their employees so they will be a valuable asset; however, the employees themselves must possess high, if not indeed exceptional, self-motivation levels to become more productive. The two groups of participants in this study verified that the four factors of self-confidence and self-efficacy, positive thinking and positive thinking about the future, focus and strong goals, and a motivating environment stimulated their self-motivation to different degrees. However, the overall results showed that these four factors affected their self-motivation levels most of the time. Thus, they revealed they are highly motivated. A high level of self-motivation is below the exceptional level of self-motivation, and hence this implies that more push factors are necessary to facilitate remarkable and unique output. These additional push factors could be personal challenges for every teacher and new activities could broaden their understanding and capabilities. Advanced studies outside the workplace can increase motivation levels: new skills and qualifications are noble weapons that boost morale and self-motivation.

Last but not least, energy booster activities are important. A person must try to avoid stress so that their self-motivation attitudes will keep on soaring rather than diminishing. Incomparably high self-motivation attitudes would lead to the attainment of their goals.

REFERENCES

Fournier, G. (2018) *Locus of Control, Psych Central.* Available at: https://psychcentral.com/encyclopedia/locus-of-control/ (accessed: 14 March 2019).

Goal Setting Tools (2014) *Self-Motivation Skills.* Available at: http://www.mindtools.com/ pages/ article/ new LDR_57.html (accessed: 25 May 2014).

Hans, T. A. (2000) A meta-analysis of the effects of adventure programming on locus of control, *Journal of Contemporary Psychotherapy.* Springer, 30, 1, 33–60.

Mamlin, N., Harris, K. R. and Case, L. P. (2001) A methodological analysis of research on locus of control and learning disabilities: Rethinking a common assumption, *The Journal of Special Education*, 34, 4, 214–225.

Neill, J. (2006) *What is the Locus of Control?* Available at: http://wilderdom.com/psychology/loc/LocusOf ControlWhatIs.html (accessed: 14 March 2019).

Siddiqui, F. (2016) *Self Motivation-Importance of Self Motivation.* Available at: http://selfimprovement tips.expertscolumn.com/article/self-motivation-importance-self-motivation (accessed: 15 March 2019).

Emerging Trends in Psychology, Law, Communication Studies, Culture, Religion, and Literature in the Global Digital Revolution – Setiawan & Rahmawati (eds)
© 2020 Taylor & Francis Group, London, ISBN 978-1-03-224216-3

Legal protection for consumers in online transactions

Dharu Triasih, B. Rini Heryanti & Endah Pujiastuti
Universitas Semarang, Semarang, Indonesia

ABSTRACT: Online e-commerce is believed to be easier and more cost-effective than conventional business. But this does not also guarantee total security in conducting transactions. The losses to consumers from online shopping also vary, ranging from products that do not match the descriptions, low quality items, to fraud. This study aimed to research the legal protection for consumers against fraudulent acts from sellers. The approach used in this research is legal empirical. Magelang, Pekalongan, and Semarang were the selected research areas, using purposive non-random sampling. The primary data was obtained by interview, and a literature review was used for secondary data. The collected data was then analyzed using qualitative methods. Based on this study it can be concluded that the implementation of online transactions is based on mutual trust between the parties. Consumers are the weak party in this online sale and purchase agreement, often experiencing loss and dissatisfaction when receiving the goods or services promised, with incorrect delivery times, inappropriate items, and even fraud. The existing law is not sufficient to provide optimal legal protection to consumers.

Keywords: legal protection, buying, and selling, online, consumers

1   INTRODUCTION

1.1   *Background*

The business of buying and selling online (e-commerce) is believed to be easier and more cost-effective than with conventional businesses. But this does not also guarantee total security in conducting transactions. At present, there are still many types of fraud and other crimes in cyberspace, known as cybercrime (Halim Barkatulah, 2008). It is common for people to feel disappointed with the goods and services obtained from shopping, both offline and online. In today's digital era, there is much opportunity, which a high potential for consumers to suffer losses.

The losses to consumers from online shopping also vary, ranging from products that do not match the descriptions, low quality items, to fraud.

The enthusiasm of the community regarding e-commerce business has not receded, which has caused a shift in shopping patterns, from conventional to virtual. This, of course, is rapidly increasing the value of e-commerce transactions year on year. In this regard, the role of the Indonesian government as a regulator is crucial, so that the country is not only able to compete in the world arena but is also able to provide legal protection to consumers in this online sale and purchase agreement. The government as a regulator needs to create rules that on one hand protect consumers, especially against global businesses, and on the other hand create a situation that is conducive to the growth of online businesses in this country.

1.2   *Research problem*

What legal protection exists for consumers in purchasing and selling online based on consumer protection and electronic transaction laws?

2 METHOD

Fundamentally, legal research is a science activity based on certain techniques, systematics and thoughts. It aims to study and analyze one or several specific legal symptoms, with a thorough examination of these legal facts.

This research used the empirical legal approach, which is a legal research method that examines not only statues, acts, legal principles, but also legal phenomenon in society.

According to legal research, the secondary data needed are as follows. 1) Primary legal material, consisting of the three books of the Indonesian Civil Code about obligations: Law of the Republic of Indonesia No. 36 of 1999 relating to Telecommunications; Law of the Republic of Indonesia No. 8 of 1999 relating to Consumer Protection Law; and Law of the Republic of Indonesia No. 11 of 2008 relating Information and Electronic Transactions. 2) Secondary legal material that provides directions and explanations on primary legal materials, such as the work of legal specialists and other records related to the issue being studied. 3) Tertiary legal material, that provides directions and explanations for secondary legal and primary law materials, i.e. legal dictionaries, magazine/journals or newspapers, insofar as data is applicable to the study material. The field research survey was performed by conducting in-depth interviews with informants and respondents, with questions determined by the scientists to achieve primary data.

3 RESULTS AND DISCUSSION

3.1 *Legal protections for online customer protection, sale and purchase agreements*

Broadly speaking, consumers who shop online or offline have the same protection of rights. However, there are several legal approaches to consumer protection in online shopping/purchasing transactions. An online transaction or shopping also refers to Government Regulation No.82/2012 concerning the Implementation of Electronic Transactions and Systems (PP PSTE), which is derived from Law No.11/2008 concerning Information and Electronic Transactions (UU ITE, 2008).

This regulation considers online transactions or shopping to be quite different from offline shopping. Albeit the laws that apply in the transaction are the same, but the media used is different. Online shopping is the use of the internet or other telecommunications facilities. Thus, consumers have the right to know what obligations must be met by the seller or manufacturer.

The obligations of sellers in an online transaction, based on Article 7 Consumer Protection Act No. 8 year 1999, are as follows:

1. Apply the principle of good faith in carrying out a business activity.
2. Provide information correctly, clearly, and honestly, with regard to the conditions and guarantees of goods/services, as well as an explanation of how to use, to repair, and to maintain.
3. Serve properly, honestly and not discriminatively.
4. Provide quality assurance of goods/services according to applicable standards.
5. Allow consumers to test and try, and also give a guarantee.
6. Give some compensations for losses due to use.
7. Give compensation and a replacement if the item is not as per the agreement.
   The most important thing that regulates consumer protection, both conventional and online purchasing and selling, is found in Article 4 of Law No. 8 of 1999 (Miru and Yodo, 2011):
a. The right to comfort, security and safety in the comfort of goods and/or services;
b. The right to choose goods and/or services and obtain the said goods and/or services following the exchange rate and the promised terms and guarantees;
c. The right to true, clear and honest information regarding the requirements and guarantee of goods and/or services;

d. The right to be accompanied and filed for the goods and/or services used;
e. The right to advocacy, protection, and efforts to resolve protection disputes by consumers in general;
f. The right to consumer guidance and education;
g. The right to refuse or deny, true and honest and not discriminatory;
h. The right to receive compensation where goods and/or services received are not following the agreement or not received properly;
i. The rights stipulated in the provisions of other laws and regulations.

Based on the Consumer Protection Act, if a business actor does not carry out obligations or breaches its obligation, then this attracts a maximum sentence of 5 years in prison or a maximum fine of 2 billion rupiahs. If the business actor/seller uses a fake identity or commits fraud in online buying and selling, this is a crime following the Criminal Code (Penal Code) about fraud and attracts a maximum sentence of 6 years in prison or a maximum fine of 1 billion rupiahs.

In this case the articles are the legal basis for both business actors, in this case conventional traders and online traders. It is important to note that business actors are prohibited from trading goods/services or services that are not suitable as stated on labels, etiquette, information, advertisements or promotion of the sale of goods and/or services, nonconformity of goods specifications of goods received with goods listed in the advertisement/photo of goods offering is a form of violation/prohibition for business actors in trading goods.

Based on Article 4 of the Consumer Protection Act, the consumer has the right to receive compensation and/or reimbursement if the goods received are not according to the agreement, and business actors following Article 7 of the Consumer Protection Act are obliged to give compensation.

Online sale and purchase agreements are increasing in Central Java, and there are several obstacles. Problems such as inequality of internet access on and outside Java and inadequate wired/wireless network infrastructures are technical obstacles for online businesses. In addition, five other factors hinder the continuity of online activities in Central Java: public awareness, security in transactions, limited internet banking facilities, and culture. These constraints should be regulated through an adequate set of policies. While there are no comprehensive policies on e-commerce in Central Java, some general rules relating to trade and the use of information technology provide a guide for residents of Central Java in running e-commerce, including several articles in the Criminal and Civil Code.

Article 362 of the Criminal Code can be used to ensnare carding actors, namely theft of credit card numbers to conduct e-commerce transactions. Article 378 of the Criminal Code can be applied to perpetrators of fraud in e-commerce activities, for example parties who display advertisements on websites to entice visitors to buy items that are not ever sent. Article 1233 of the Civil Code concerning conventional buying and selling has very different characteristics, so that comprehensive rules regarding e-commerce in Indonesia are necessary. There are currently no laws that specifically regulate basic rules and technical infrastructure that support security and realization of e-commerce in Indonesia.

Act No. 11 of 2008 about Electronic Information and Transactions was initially expected to answer all public issues related to IT development in Indonesia. Electronic transactions are specifically discussed in Chapter V of the Act, namely from Article 17 to 22. However, in reality the law is not sufficient to become a legal umbrella in the activities of electronic transactions, especially in online sales and purchases.

Another legal problem is the occurrence of fraud by businesses when buying and selling activities online. Online transactions are new buying and selling activities that utilize technological advancements. Online transactions are attracting increasing attention from enthusiasts of buying and selling online along, with technological developments that facilitate the buying and selling process. This is caused by people's needs for fast, easy and practical services and the community now has a wider space to choose products. The high level of complaints by consumers in Indonesia related to fraud when buying and selling online certainly requires attention. This means that consumers require legal protection for when problems occur.

Online banking is more practical and easy and can be done at any time with an internet connection, but conversely can have a negative impact, namely the emergence of legal problems that can cause losses to consumers. The possibility of fraud is also prevalent, caused by a lack of information that is often accepted by consumers. The validity of the transaction process has been explained in the Civil Code (KUHP, 1847) Article 1458 which states that sale and purchase is considered to have occurred between the two parties as soon as they reach an agreement on the item and its price, even though the item has not been delivered and the price has not been paid.

Therefore, there is a need for legal protection for consumers who conduct online transactions, especially because consumers have universal rights that must be protected, namely the right to security and safety and the right to the right information. In Indonesia there are currently no specific laws governing online transactions. Likewise, Law Number 11 of 2008 concerning Information and Electronic Transactions cannot yet be used as a basis for handling fraud cases in online transactions.

The current law in Indonesia that can be used as a guideline in this matter is Law Number 8 of 1999 about Consumer Protection (UUPK, 1999), which aims to create a consumer protection system that consist elements of legal certainty and information disclosure and access to information, even though it does not specifically regulate online transactions. The following Articles can be used as guidelines in resolving fraud cases in online transactions:

1. Article 8 paragraph (1) letters d, e, and f, which states that business actors are prohibited from producing and/or trading goods and/or services that are not in accordance with quality, conditions and promises as stated on labels, information, advertisements or promotions for the sale of said goods and/or services.
2. Article 16 letters a and b, which state that business actors keeping orders and/or agreement on the agreed time of settlement and are prohibited from keeping promises of services and/or achievements.

Thus, it can be concluded that there is a need for legal certainty for protection for customers who make online transactions. This is because customers have important rights that must be upheld, and to foster awareness of business actors regarding the importance of customer protection so that an honest and responsible attitude in the business grows. In this case, besides the existence of UUPK, regulations specifically regulating online purchasing and selling activities are needed because these not only protect customers but also online businesses.

4 CONCLUSION

Online sale and purchase agreements concerning legal protections for consumers in Central Java is prioritized to the trustworthiness of parties aspects. Consumers are the weak party in this online sale and purchase agreement, often experiencing loss and dissatisfaction when receiving the goods or services promised, with incorrect delivery times, inappropriate items, and even fraud. The existing law is not sufficient to provide optimal legal protection to consumers.

REFERENCES

Halim Barkatulah, A. (2008) Hukum Perlindungan Konsumen. Kajian Teoretis dan Perkembangan Pemikiran, Nusa Media, Bandung.
KUHP (1847) *Kitab Undang-Undang Hukum Perdata.*
Miru, A. and Yodo, S. (2011) Hukum Perlindungan Konsumen, cetakan ke 7. *PT Rajagrafindo Persada, Jakarta.*
UU ITE (2008) *Undang-undang Informasi dan Transaksi Elektronik.*
UUPK (1999) *Undang-Undang Perlindungan Konsumen.*

Emerging Trends in Psychology, Law, Communication Studies, Culture, Religion, and Literature in the Global Digital Revolution – Setiawan & Rahmawati (eds)
© 2020 Taylor & Francis Group, London, ISBN 978-1-03-224216-3

Legal protection of passing-off well-known marks for the welfare of consumers in Indonesia

Doddy Kridasaksana
Universitas Semarang, Semarang, Indonesia

ABSTRACT: The success and good reputation of well-known marks in the marketplace will undoubtedly sometimes make business actors operate in bad faith, one of which is conducting passing-off actions. Consequently, this can harm consumers, culminating in an absence of prosperity, security, and safety. Therefore, there is a need for legal protection.The objectives of this study are as follows: 1) to analyze the legal protection in Indonesia against passing off the well-known marks under Act Number 20 the Year 2016 on Trademark and Geographical Indication; 2) to analyze protection for consumer welfare of passing off acts upon the well-known marks in Indonesia. Legal protection in Indonesia consists of preventive and repressive actions. The state should protect the welfare of consumers to accommodate all interests in order to create legal protections so that consumers can feel prosperous, comfortable, and safe. This is different to previous research because of the legal protection of passing off in Indonesia based on Law Number 20 the Year 2016 on Mark and Geographical Indications and Law Number 8 the Year 1999 on Consumer Protection.

Keywords: legal Protection, passing Off, well-known Mark, consumer

1   INTRODUCTION

1.1   *Background*

One of the elements of intellectual property rights related to trading must be protected is a mark. We can use the mark as a means to stimulate the growth of the industry, and ensure trading that is fair and profitable for all parties, especially for well-known brands. If a product does not have a strong brand then it will not be known by consumers or the public (Wikipedia, 2017).

The success and the good reputation of a brand, especially well-known marks in the marketplace, will inevitably tend to ensure other producers or entrepreneurs try to compete. We conduct this through bad faith (one of the examples ispassing off), which impacts on consumers who are aggrieve. Moreover, legal uncertainty can protect them (Syarifin and Jubaedah, 2004).

1.2   *Objectives*

The objectives of this study are:

1. To analyze the legal protection toward passing-off actions over well-known marks under Law Number 20 the Year 2016 on Mark and Geographical Indications as well as such cases in Indonesia.
2. To analyze the protection for consumer welfare toward passing-off actions overwell-known marks in Indonesia.

2 LITERATURE REVIEW

2.1 *Legal protection and consumer protection*

Legal protection is an effort to achieve or create a code of conduct, security and peace in society, whether this is with defensive or repressive efforts. Legal protection is all government efforts to ensure the availability of legal certainties to citizens and to place sanctions on the violators by the applicable regulations (Maulana, 1999).

2.2 *The definition of mark, well-known mark and passing off*

By the understanding of Article 1 (1) of Law Number 20 Year 2016 (Hukum Online, 2017), a mark can be displayed graphically in the form of a picture, logo, name, word, letter, number, color arrangement, in two dimensions or three dimensions, sounds, holograms, or a combination of two or more of these elements. They distinguish goods or services produced by a person or legal entity in the goods or service trade activities.

Converting a mark into a well-known mark is not easy, takes a long time, and is costly. A mark can become well-known because of intense advertising, installed in various mass media, and the good quality of the goods, for example, the Coca Cola soft drink brand from the United States.

The definition of passing off according to Black's Law Dictionary (Garner, 2004) is the act or an instance of falsely representing one's product as that of another in an attempt to deceive potential buyers. Passing off actionable in tort under the law of unfair competition. It may also be actionable as trademark infringement. Even there is one who defines it as an action that seeks to gain profit through shortcuts by all means and the pretext of violating business ethics, norms of decency and law (Djumhana and Djubaedillah, 2014).

2.3 *Consumer protection*

In the era of globalization and free trade, there are many kinds of goods/service products marketed to consumers in Indonesia, either through promotion, advertising, or supply of goods directly. If they are not careful in choosing their desired product/service, consumers risk becoming the object of exploitation byan irresponsible business actor. The Government realizes that legislation and regulations are required in all sectors related to the movement of goods and services from employers to consumers. The government also must oversee the enactment of the legislation and regulations well through Law Number 8 Year 1999 on Consumer Protection (Radio prrsni, 2017).

3 DISCUSSION

3.1 *Legal protection toward passing off actions*

Protection of well-known marks is preventive legal protection and repressive legal protection. Preventive legal protection is protection before criminal offence or violation of law toward a brand and well-known mark. If a person/legal entity wishes legal protection for its brand under the law, the relevant brand must be registered. The vital requirement that hall marks a mark is the existence of sufficient distinctive power (distinctiveness). Repressive legal protection toward marks occurs when there is criminal offence or violation of the right to the mark. This repressive legal protection is granted if there is a violation of the rights over the mark (including well-known marks). In this case, the role of judicial institutions and other law enforcement agencies such as the police, civil servant investigator (CSI), and prosecutor are crucial.

So far the legal protection toward well-known marks refers to the provisions of Article 6 of the Paris Convention and Article 16 of the TRIPs Agreement. These provisions have been adopted by many countries, including Indonesia, in their national legislation, especially in the field of marks (Syafrinaldi, 2006).

The law regulating passing off has not existed until now because Indonesia adheres to the Civil Law legal system (S. Man). The Directorate General of HKI only handles passing off cases indicated by a breach of a mark such as equality in essence or whole (using a particular form, display/packaging or design/logo), for example the case of "Oreo" (PT Kraft Food) versus "Rodeo" (PT Nissin Biscuit) (Kompas.com, 2017). The applicant registered the "Rodeo" Brand at the Brand Office only as a word using block capital letters, but in the market place the name was used in the same writing and packaging form "OREO" for the same type of goods. Similar or alike packaging can deceive or mislead consumers who are not too aware of the trade product. Besides, consumers can think that Oreo and Rodeo come from the same source; in fact, both products have different qualities because they come from different sources (Mahasiswa.me, 2017).

3.2 Protection for consumer prosperity toward passing off action over well-known mark in Indonesia

The main goal in the business world is to seek profit (Prasetijo and JoiIhalauw, 2006). Entrepreneurs who see it as one of the business opportunities will then seek to gain an advantage through an unfeasible shortcut by producing or marketing goods or products by forging, imitating, piggy backing on the well-known marks. For the consumers, this may bring distinct prestige when they wear or use the well-known mark. The use of a well-known unauthorized mark is qualified as a disgraceful use of the mark (Anneahira, 2017).

The weakness of the consumers' position means consumer protection law is essential. As a form of protection for consumers, Law Number 8 Year 1999 on Consumer Protection was established. Article 1 (1) of the Law says that consumer protection is an effort that ensures legal certainty to protect consumers. Consumer protection itself aims to raise awareness by business actors of the importance of consumer protection to grow honest and responsible attitudes in business.

There are different perceptions within society or consumers concerning brands, which creates various interpretations; nevertheless, the actions of business actors in producing an item by piggybacking the fame of others cannot be justified. Allowing the irresponsible actions will indirectly result in and justify cheating and dishonesty. The act of using another well-known mark, as a whole, does not only harm the owner or the holder of the brand itself and the consumers, but the broader impact is detrimental to the national economy and, more widely, to international economic relations. This will have an impact on the absence of the welfare, security, comfort and safety of the community or consumers. They deserve to be protected as written in Law Number 8 of 1999 on Consumer Protection, article 4 point (a), (b) and (c) on consumer rights. Business actors also have an obligation as written in article 7 point (a), to (g).

4 CONCLUSION AND SUGGESTION

4.1 Conclusion

1. The law that regulates explicitly passing off in Indonesia up to now has not existed. We can only categorize the passing off action as trademark infringement.
2. The state regulates the protection for consumers welfare in Law Number 8 Year 1999 on Consumer Protection, and the purpose is to create legal protection so that consumers can feel prosperous, comfortable, and safe.

4.2 Suggestion

1. Should have a strategic role in the field of legislation in order to provide legal certainty to the well-known mark infringement action, including passing off.
2. The awareness of the business actors about the importance of consumer protection needs to expand, and from the consumer side, the sense of loving the domestic products and proud to use it should improve too.

REFERENCES

Anneahira (2017) *Indonesia dan Pembajakan*. Available at: http://anneahira.com/indonesia-dan-pembaja kan.html (Accessed: 6 June 2017).

Djumhana, M., Djubaedillah, R. and Hak Milik Intelektual, S. (2014) Teori dan Praktiknya di Indo-nesia, *Citra Aditya Bakti, Bandung*.

Garner, B.A. (2004) Black's Law Dictionary, Eight Edition, *West Group*.

Gunawati, Anne. 2015. PerlindunganMerekTerkenalBarang dan JasaTidakSejenisterhadapPersaingan Usaha TidakSehat. Penerbit Alumni. Bandung.

Kompas.com (2017) *Passing off, Modus Baru Pelanggaran Merek*. Available at: http://nasional.kompas. com/read/2008/09/09/20040774/passing.off.modus.baru.pelanggaran.merek (Accessed: 5 June 2017).

Mahasiswa.me (2017) *Sengketa Oreo vs Rodeo*. Available at: https://mahasiswa.me/2017/05/06/sengketa-oreo-vs-rodeo/ (Accessed: 5 June 2017).

Maulana, I.B. (1999) *Perlindungan merek terkenal di Indonesia dari masa ke masa*. Citra Aditya Bakti.

Prasetijo, T. and JoiIhalauw, J. (2006) *Perilaku Konsumen*. Jogjakarta: Andi Publisher.

Syafrinaldi (2006) *Hak Milik Intelektual dan Globallisasi*. Ulir Press: Pekanbaru.

Syarifin, P. and Jubaedah, D. (2004) *Peraturan Hak Kekayaan Intelektual di Indonesia*. Pustaka Bani Quraisy.

Ustman, Sabian. 2010. MenujuPenegakanHukumResponsif. PustakaPelajar. Jogjakarta.

Wikipedia (2017) *Mark on trademark is a name or symbol associated with goods/services and arousing psy-chological/association meaning*.

http://semestahukum.blogspot.co.id/2017/02/perlindungan-konsumen-terhadap.html. Accessed on 4 June 2017.

http://www.radioprssni.com/prssninew/internallink/legal/uu_8_99perlkonsum.htm. Accessed on 5 June 2017.

ww.hukumonline.com/pusatdata/downloadfile/.../lt5850fd588be8. UU RI No 20 Tahun 2016 tentangMerek-danIndikasiGeografis. Accessed on 6 June 2017.www.

http://agungsujatmiko73.blogspot.co.id/2011/11/penguatan-prinsip-kebebasan-berkontrak.html

Resilience in adolescents infected with HIV/AIDS

Anna Dian Savitri & Purwaningtyastuti
Universitas Semarang, Semarang, Indonesia

ABSTRACT: Research has aimed to describe the resilience of adolescents infected with HIV/AIDS. The subjects in this study are teenagers positively infected with HIV/AIDS, who are undergoing ARV treatment, and who are willing to be the subject of research. Informants who supported this study were volunteers and leaders of shelter houses where the subjects resided. The method used in this study was a qualitative research method with a case study approach. Subject and informant data was collected through structured interviews, and the analysis technique used source triangulation by comparing the analysis of subject data with research informants. The results of this study are that subject has a high level of resilience. The aspects I Have includes support and attention from other people; the subject prioritizes pleasure and comfort, has role models, has the drive to be independent, and has experienced health discrimination, has received a good education and The factor I Am includes having an attractive attitude, expressing affection through actions, and caring. The subject refrains from transmitting HIV/AIDS, is independent and responsible, proud of herself, and will stay healthy, have life expectancies and be sure to achieve them. The aspects I Can includes being able to express what is felt and thought, to solve problems, to control emotions, and build good relations with others. Based on the results of the study it can be concluded that the subject has good resilience and has been able to accept h condition.

Keywords: Resilience, Adolescence, HIV/AIDS

1 INTRODUCTION

In adolescents who are infected with HIV/AIDS (called ODHA, people with HIV AIDS), if they have resilience they will avoid a variety of negative risks that interfere with their psychology, and at least be able to recover and adapt positively like teenagers in general. The subject was a young woman (age 18 years) who was positively infected with HIV AIDS. In an interview conducted by researchers in December 2017, the subject explained that when she was diagnosed with HIV/AIDS she found it very difficult psychologically. But gradually she tried to wake up from her nightmare in order to avoid the worst conditions and stress, even though at certain moments she still felt inferior due to her condition. The ability to manage stress well helps her think positively even though she is faced with pressure from the surrounding environment, and this develops optimally through doing activities in a foundation or NGO together with some people who have the same diagnosis. The subject did not feel prolonged sadness. Her tough life after receiving the HIV/AIDS diagnosis, made her stronger for coping with various problems Bad experiences succeeded in making her a strong teenager, especially the poor economic situation of her family.

Luthans (2005: 325) suggests that a person's resilience can be developed by increasing their assets through education, training, and by maintaining social relations and generally improving the quality of the available resources. Based on the above data it can be concluded that resilience is necessary, especially for adolescents who are infected with HIV/AIDS. Individuals who are able to deal with stress, depression, and difficult conditions, have reasons and goals to fight for their lives. Resilience is a capability that is very much needed in the lives of every person, group or society. It enables them to face, adapt and overcome life's difficulties and challenges through long and complicated processes, where competencies are developed whenever interacting with the daily environment in a positive way and helps eliminate the detrimental impacts.

Grotberg (Desmita, 2013: 229) mentions the characteristics of individual resilience is having:

> I have is a source of resilience that is related to the meaning of someone to the amount of support given by the social environment to her. I am is a source of resilience which is related to the personal power a person has, and consists of feelings, personal attitudes, and beliefs. I can is a source of resilience related to what can be done by someone in connection with skills.

According to East Java Regional Regulation (2004), people with HIV and AIDS are those who have been infected with HIV both at the symptomatic and symptomatic stages. To identify people infected with HIV, you can test directly for the HIV virus or indirectly by finding antibodies. If someone has antibodies to HIV, that person has been infected with HIV (MOH, 2003). People are said to be infected with HIV if they have taken a voluntary counseling and testing test (VCT) without coercion. This examination is a frequent one, carried out three times; if the first and second results are negative, it is possible for the virus to still be in the window period or, the period when the serologic test for HIV antibodies still shows negative results while the virus is present in large amounts in the blood of the sufferer. At the third test, if the result is still negative then the person does not have HIV virus but if the result is positive then there is HIV virus present (Yoga, 2010).

2 RESEARCH METHODS

Subjects of the study were selected using purposive methods, based on the objectives of this study. Sugiyono (2007: 219) explains the purposive is a sampling technique of data sources with certain considerations, for example the person is considered to know best about what is expected or maybe he is the ruler so that it will be easier for researchers to explore the object or social situation under study.

The subject is a 19-year-old student at the Open Middle School in Semarang. She was infected with HIV/AIDS through sexual relations due to being a victim of trafficking Now she continues to communicate and socialize with her friends and not shut herself down. She believes that even though she is infected with HIV/AIDS her body can still be fresh like a healthy person without the disease. She has became one of the activists at the Rumah Aira Foundation. This study used a case study method to reveal the picture of self-resilience and the factors behind it. In addition, the selection of this method is based on the fact that the themes in this study are unique.

3 DISCUSSION

Being infected with HIV/AIDS will cause changes in experience such as behavior and feelings. The dynamics of the subject before and after accepting their condition will be different every time. The subject has a variety of severe problems: at a young age she had to leave school and became a victim of trafficking, and had to have sex against her will. When she was diagnosed as HIV positive, she was silent and without words, unable to cry and scream, as if she did not believe it. At that time the subject was worried whether there would be people who would still accept her, but she had people close to support her even though they already knew about the health condition. This made her feel safe and comfortable and she became stronger because the people closest to her were always there. When taken by Social Services to the Rumah Aira Foundation, which specifically accommodated PLWHA, she learned from the experiences of other people with HIV/ AIDS, which made her stronger because she felt not alone. In addition, she also received good health, education and security services. She is loved by others because she is friendly, cheerful, sociable, socially high and liked the new environment. She was also able to show her affection to others, by caring for her younger sister in the foundation. In addition, she is a person who cares about other people and helps them refrain from transmitting HIV/AIDS to others, making other people more willing to help themselves because they consider her to be admirable.

The subject is now is one of the activists in the Foundation, providing peer support. Despite her positive HIV/AIDS diagnosis, she is proud because she can help other people with HIV/AIDS and the economic state of their families. This makes the assessment of her more positive. When experiencing problems, trust and self-esteem can help the subject survive and overcome existing problems. Her hopes for the future and beliefs that she possesses further increase her chances to rise from adversity.

4 CONCLUSION

Based on the results of the research conducted the description of the resilience of the subject can be seen from the aspects of I Have, I Am, and I Can.

I Have includes getting support and attention from other people, and subjects are more concerned with pleasure and comfort, have role models, have an incentive to be independent, receive good education and security services. In I Am the subject has an interesting nature and a feeling of being loved by others, is able to express affection through deeds and caring for others, and refrains from transmitting HIV AIDS to other people. The subjects are independent and responsible even though they are not maximal, they will stay healthy, have a good life expectancy, and are able to make that happen. In I Can, the subject is able to express what they feel and think, solve the problem at hand, find the help they need and establish good relations with others. The subject was infected with HIV/AIDS because she was a victim of trafficking.

REFERENCES

AIDS Commission (KPA). 2003. *National AIDS Strategy, 2003–2007*. Menkosesra, KPAN, Jakarta.

Azwar, S. 2003. *Preparation of the Psychological Scale*. Yogyakarta: Offset Student Library.

Calhoun, J.F. and Acocella, J.R. 1995. *Psychology of Adjustment and Human Relations*. New York: Mc Graw Hill.

Chaplin, C.B. 1995. *Complete Dictionary of Psychology*. (Translation: Kartini Kartono). 1st edition, Jakarta: Grafindo Persada.

Davidoff. 1991. *An Introduction to Psychology*. Volume 2. Interpreting: Mari Jumiati. Jakarta: Erlangga.

Gerungan, W.A. 1996. *Social Psychology*. Bandung: PT Eresco.

Herdiansyah, H. 2015. *Qualitative Research Methodology for Psychology*. Jakarta: Salemba Humanika.

Hurlock, E. 2004. Developmental Psychology. Jakarta: Erlangga Press.

Ministry of Health of the Republic of Indonesia. 2012. MEDIAKOM. *Getting to Know, Preventing Growth of HIV-AIDS*. Jakarta.

Sarafino. 1998. *Health Psychology: Biopsychosocial Interaction*. USA: John Willey and Sons.

Smet, B. 1994, *Health Psychology*. Jakarta: PT Grasindo.

Emerging Trends in Psychology, Law, Communication Studies, Culture, Religion, and Literature in the Global Digital Revolution – Setiawan & Rahmawati (eds)
© 2020 Taylor & Francis Group, London, ISBN 978-1-03-224216-3

Reconstruction of the legal protection concept Janda Cerai in the Batak Toba community based on the value of justice

Amri Panahatan Sihotang
Universitas Semarang, Semarang, Indonesia

ABSTRACT: Customary law is a living law that grows and develops in the midst of society in accordance with the development of that society. Batak society is a patriarchal one with a patrilineal kinship system. The kinship system in a patrilineal society also affects the status of widows and daughters. The position of widows according to adat is based on the principle that women as foreigners are not entitled to inherit. The patrilineal community contains the provision that if widows are integrated into their husbands' relatives' families, the widow can settle in that community and earn a living. However, if the widow separates herself from her husband's relatives by remarrying outside of her husband's relatives, the widow will never inherit the property of her husband. Boys have dominance over girls and are very influential on the status of divorced widows in Batak society. The aforementioned provisions become customary law applicable wherever the Batak Toba community is located and are a form of injustice toward women and divorced widows in the Batak Toba community. If the legal settlement is done through a formal law, it would clearly be in conflict with customary values. Through the constructivism paradigm and sociolegal research approach, the author aimed to thoroughly examine the issue of protection of women and divorced widows within the Toba Batak customary law.. The benefits of this research include providing a means to generate new ideas related to the implementation of customary law development in a gender perspective. The patriarchal culture sis so strongly embraced by the Batak Toba community that it gave birth to the normativity that divorced widows are placed in a lower position than those who are not widows. A divorced widow is afforded protection through the renewal of customary law, the Treasure, which gives her the authority to manage the inheritance of her husband without any interference from the husband's relatives as long as the she does not remarry. Joint Treasure gives the widow the power to own the estate without any interference from the family or relatives of the husband even if she remarries. For protection of divorced women on the death of their husbands on the basis of the certainty of customary law, the government should immediately conduct a judicial revision of Law No. 1 of 1974 to clarify the status and rights of divorced widows . Social engineering in the adat customary law as related to the protection of divorced widows is based on the premise that elders should practice tolerance toward divorced widows to determine their lives after the death of their husbands without being shackled by an "honest" or sinamot marriage imposed by men to women or relatives.

Keywords: divorce widow, justice value, legal protection, reconstruction

1 INTRODUCTION

In Indonesia customary law is an unwritten rule that lives within the indigenous peoples of a region and will remain alive as long as the people still adhere to the customary laws that have been passed on to them from their ancestors. Therefore, the existence of customary law and its position in the national legal system cannot be denied even though customary law is

29

not written and does not have legal status. Customary law will always exist and live in the community and is still needed to answer the demands posed by the complexity of globalization issues. It addresses the demands of society, which are truth and justice, not procedural law.

Talking about customary law is fraught with social problems. First, the concept of customary law that was invented by C. Van Vollen Hoven and developed by universities is certainly not relevant in the present. According to C. Van Vollen Hoven, customary law is the rule of law applicable to indigenous people and foreign easterners, who on the one hand have sanctions (hence the term "law") and on the other hand have codified the law (hence the term "custom"). Second, customary law that is the source of national law has not been legalized as a written rule, and in the context of local culture needs to be developed in the present era so that the law in the community feels more inherent, acceptable, and adaptive. Third, the settlement of adat disputes does not recognize the separation between criminal and civil cases. Fourth is the horizontal separation of the land law.

1.1 *Problem formula*

Based on the background outlined in the preceding section, in this study the issues to be studied were formulated as follows:

1. Why is the position of the divorced widow's in Batak Toba society considered to be justified?
2. Which legal protection is afforded divorced widows according to Islamic law, marriage law, Christian law, and customary law outside Toba Batak Society?
3. How is the reconstruction of the detachable widow law concept ideal for Batak Toba society based on the value of justice?

2 RECONSTRUCTION OF THE LEGAL PROTECTION CONCEPT OF THE IDEAL DIVORCE DORMAGE IN THE BATAK TOBA COMMUNITY BASED ON THE VALUE OF JUSTICE

2.1 *Protection of divorced widows law through the reform of customary law*

Although the customary rules prevailing in this Batak Toba society are already undergoing changes, especially in urban areas, they have not had significant effects. The "honest" marriage system (sinamot), as part of the patrilineal system, still persists, wherein the inheritance of most of the family is still based on customary law and its granting of mercy from relatives of men or boys of the family. Divorced widows and women have no power in terms of custom or decision-making in other social activities in the Batak Toba community (Interview with Ny. Hutabarat boru Panjaitan, Semarang, August 14, 2016).

The author attempts to provide a discourse on the reconstruction of the concept of protection of divorced widows in Batak society in terms of the following aspects of inheritance:

a. Luggage, which is the property brought by the husband before marriage where the divorced widow has the authority to manage the inheritance of her husband without any interference from the husband's relatives as long as the widow does not remarry.
b. Joint Treasure, which is the property acquired jointly during the marriage period in which the divorced widow is given the power to own the inheritance without any interference from the family or relatives of the husband even if the widow remarries. This is consistent with the theory put forward by Radbruch that teaches that we must use the priority principle in achieving the objectives of the law, and the first priority always falls on justice, then the benefit, and further legal certainty. An attempt is made to apply the two concepts in the same way to divorced widows whose husbands died in Bonapasogit (hometown) and divorced widows whose husbands died overseas or in urban areas.

2.2 *The protection of the divorced widows law refers to the certainty of customary law*

If we were more inclined to stick to the value of legal certainty or from the point of the rules, then it would immediately shift the values of justice and usability, because what is important to the value of certainty is the rule itself. Whether the regulation has fulfilled a sense of justice and usefulness to society is beyond the priority of the value of legal certainty. Similarly, if we are more likely to stick to usability values, then it will shift the value of legal certainty as well as the value of justice, because what is important for the value of usability is whether the law is beneficial or useful to the community. Similarly, if we only hold the value of justice, then it will shift the value of certainty and usefulness, because the value of justice is not bound to legal certainty or the value of usability, as something that feels fair is not necessarily in accordance with the value of usefulness and legal certainty (hal. 29). Thus we should be able to make a comparison between the three values or be able to make a compromise that is proportionally harmonious, balanced, and aligned among them. Law is a means of social reform or a tool of social engineering. With this theory, Roscoe Pound goes on to explain that law as a social institution can be perfected through human intellectual efforts and considers it people's duty to find the best ways to advance and direct the effort. In this case the recommendation offered by the author is that the government should immediately conduct a judicial revision of Law No. 1 of 1974 (Undang-Undang Nomor 1 tahun 1974) to clarify the rights and status of divorced widows so that the customary law applicable in the community can refer to the Act.

2.3 *Social engineering in indigenous customary law is related to the protection of divorced widows according to the value of justice*

Today Indonesia has grown to understand the functioning of law as a tool of social engineering, especially in the field of transforming private adat law into a national private law (Satjipto Rahardjo, hal. 111). One aspect of the law is the creation of certainty in the relations between people in society. Closely related to the issue of certainty is the issue of the source of the law. Certainty about the origin or source of the law becomes important as law becomes an increasingly formal institution.

The customary law of the Batak Toba community embraces the partial chat system, specifying that all the property accrued during marriage belongs to the husband and the widow has no right to inherit the estate of her husband. The position of a widow in customary law is deemed incompatible with the sense of justice, and therefore widows should be given a proper position in addition to that of the children in terms of inheritance (Mahkamah Agung, November 2, 1960). The Toba Batak community kinship system embraces a patrilineal system in which after the husband of a divorced woman dies, his wife and his children become the responsibility of the relatives of her husband. However, this is fading because at present every family in the Batak Toba community has its own life and responsibilities irrespective of the interests and responsibilities that once prevailed in the patrilineal system (Interview with Ny. Sitinjak br Simamora, Semarang, January 16, 2017).

Reconstruction of the law can be done if a problem occurs in the community and there is no provision that can be executed to solve the problem, despite legal interpretation, as well as after a resolution is sought in customary law but no rules can bring a settlement to the case. In such cases, the government must reexamine the legal system that concerns the particular legal situation. If some of the provisions are held in common, then the government should make a draft law in accordance with the interests of women and justice that will be obtained by women and widows in the patrilineal community system. According to van Apeldoorn, in this case the government must adjust (waarderen) the law in line with the concrete things that occur in society and the government can add (aanvullen) to the law if necessary. The state or government in this case must adapt the law with concrete matters, because the law should cover any incidents that arise in society (E. Utrecht, hal. 230).

In reality the social life of the divorced widow in Toba Batak society has not fully realized the principle of equality between men and women.. This is because the state law relating to the issue of women and divorced widows is generally ambiguous and contradictory, and contrary to what is desired by women and divorced widows of Batak Toba society will fulfill their rights in people's lives. On the one hand the state law has provided an opportunity for equality between men and women. On the other hand, there is also the formulation of laws and regulations that lead to discrimination against women and divorced widows when the state places women in the domestic sphere, dependent on the husband, and does not protect women from violence (especially in the household). Discrimination against divorced widows and women in general indicates that unequal husband–wife power is still occurring within the Toba Batak community. In this case the adat elders should provide tolerance to divorced widows to determine their lives after their husbands have died without being shackled by "honest" or sinamot marriages that men have imposed on a woman or a relative.

3 CONCLUSIONS

Reconstruction of the ideal concept of legal protection of divorced widows for the future in the Batak Toba community should be based on the value of justice, namely:

a. The protection of divorced widows through the reformation of customary law is achieved by legal assimilation, that is, parallel legal culture is expected to give the cultural color of patrilineal law through a customary institution with the aim of:
   1. Luggage, which is the property brought by the husband before marriage where the divorced widow has the authority to manage the inheritance of her husband without any interference from the husband's relatives as long as the widow does not remarry.
   2. Joint Treasure, which is the property acquired jointly during the marriage in which the divorced widow is given the power to own the inheritance without any interference from the family or relatives of the husband even if she remarries.

b. The protection of divorced widows with respect to legal certainty can be done by conducting a judicial review of Law No. 1 of 1974 in order to establish the rights and status of the divorced widow more clearly. With the clarity and certainty of this law it is expected that customary law applicable in society can refer to the marriage law.

c. Social engineering with respect to the protection of divorced widows in customary law can be done through customary elders, providing tolerance to divorced widows to determine their lives after the death of their husband without being bound by the "honest" marriages that men have imposed on a woman or relative.

4 SUGGESTIONS

1. The status of divorced widows in Toba Batak society, which until now is still "wonder" (the feeling of fear on the part of the unfaithful widow, violating the philosophy of adat, feeling ostracized, the concept of life view semati, etc.) in Sinamot traditions related to marriage should instead be one of "Hasangapon" (praised, honorable, wise, and compassionate).
2. The position of widowed women as single parents or heads of households is also not fully recognized by the state and patriarchal Indonesian society. Law No.1 of 1974 on Marriage, for example, only mentions the wife as a housewife. This has implications for unequal rights and obligations in social life. Therefore, it is necessary to revise Law No. 1 of 1974 on Marriage so that the rights and obligations between husband and wife are the same without removing the elements that exist in customary law (adoption).

BIBLIOGRAPHY

Undang-Undang Nomor 1 tahun 1974. *Perkawinan*.

Akbar, Rizal. 2005. Tanah Ulayat dan Keberadaan Masyarakat Adat, Pekanbaru: LPNU Press.

Anderson, JND. 1959. Islamic Law in the Modern World/New York.

Aveldoorn, van L. J. Pengantar Ilmu Hukum. Jakarta: Pradnya Pramita,1986.

Azhar, Ahmad. 1994. Hukum Waris Islam. Yogyakarta: Ekonisia.

Azhari, Tahir. 2001. Karakteristik Hukum Kewarisan Islam Dalam Bunga Rampai Hukum Islam. Jakarta: Sinar Grafika.

Bahar, Syafroedin. 2006. "Upaya Perlindungan terhadap Eksistensi Hak -hak Tradisional Masyarakat Adat dalam Perspektif Hak Asasi Manusia", dalam Suwarto (dkk), mengangkat Keberadan Hak -hak Tradisonal: Masyarakat Adat Rumpun Melayu Se- Sumatera. Pekanbaru: Unri Press.

Bastian Tafal, B. 1983. Pengangkatan Anak Menurut Hukum Adat, serta Akibatnya Dikemudian Hari, Penerbit C. V. Rajawalil. Bemelen, Sita van (ed.). 1992. Women and Mediation in Indonesia. Leiden: KITLV Press.

Black, Donald. 1976. Behavior of Law. New York: Academic Press.

C. V Wallerstein, Immanuel. 1997. Lintas Batas Ilmu Sosial. Yogyakarta: Terbitan L Kis.

Cholid Narbuko dan H. 2002. Abu Achmadi. Metodologi Penelitian. PT. Bumi Aksara. Jakarta.

Dahrendorf, Ral. 1986. Konflik dan Konflik Dalam Masyarakat Industri: Sebuah Analisa Kritik, Jakarta: CV. Rajawali Dharmayua.

Daud Manurung. 1977/1988. Sejarah Kebangkitan Nasional Daerah Sumatera Utara. Jakarta: Dep. Dik.Bud.

Djohan Makmur. 1993. Sejarah Pendidikan di Indonesia Zaman Penjajahan. Jakarta: Dep.Dik.Bud.

Djoko Suryo. 1996. Pendidikan, Diferensiasi Kerja, dan Pluralisme Sosial: Dinamika Sosial-Ekonomi 1900–1990. Makalah disampaikan pada Kongres Sejarah Nasional Indonesia pada tanggal, November 12 –15.

Fauzie Rijal (ed.). 1993. Dinamika gerakan Perempuan di Indonesia. Yogyakarta: Tiara Wacana.

Fisher, J. E. & W. O'Donohue (eds.). 2006. Practitioner's Guide to Evidence Based Psychotherapy. New York: Springer.

Friedman, Lawrence. 1984. American Law. London: W. W. Norton.

Gardiner, Juliet (ed.). 1988. What Is History Today? London: Macmillan Education.

Haar. B. Ter. 1993. Beginselen en stelsel van het adatrecht, dalam buku Hilman Hadikusuma, Hukum waris Adat, Citra Aditya Bakti, Bandung.

Hadikusuma. Hilman, 1994. Hukum waris Adat, Citra Aditya Bakti, Bandung.

_____, 1994. Hukum Perkawinan Indonesia Menurut Perundangan.

Halim A. Ridwan, S.H. 1989. Hukum Adat Dalam Tanya Jawab, Jakarta.

_____, 2006, Kebudayaan Bugis, Makassar: Dinas Kebudayaan dan Pariwisata Provinsi Sulawesi.

Selatan.Hazairin. 1967. Hukum Kewarisan Bilateral menurut Al-Quran, Tinta Mas. Jakarta. Hommes, Van Eikema, Logika en Rechtsvinding, Tanpa kota: Vrije Universiteit, tanpa tahun. Hukum Adat Hukum Agama, Mandar Maju, Bandung.

_____. Pokok-pokok Pengertian Hukum Adat, Penerbit Alumni, Bandung 1980.

Made, Suathawa. 2001. Desa Adat: Kesatuan Masyarakat Hukum Adat di Propinsi Bali, Bali: Upada Sastra. Direktorat Pemberdayaan Komunitas Adat Terpencil, 2003, Atlas Nasional Persebaran Komunitas Adat Terpencil. Jakarta: DitjenPemberdayaan Sosial Depsos RI.

LAM Jambi. 2005. "Fakta dan Pengalaman Lembaga Adat Propinsi Jambi dalam Memperjuangkan Hak Tanah Ulayat Masyarkat Adat Jambi", dalam Rizal Akbar (dkk), Tanah Ulayat dan Keberadaan Masyarakat Adat, Pekanbaru: LPNU Press.

Lie, Shirley. 2005. Pembebasan Tubuh Perempuan: Gugatan Etis Simone de Beauvoir terhadap Budaya Patriarkat. Jakarta: Grasindo.

Little, Daniel. 1991. Varieties Social Explanation: An Introduction to the Philosophy of Social Science. San Francisco: Westview Press.

Lubis, M. Solly. 1989. Serba-serbi Politik dan Hukum, Bandung: Mandar Maju.

Manurung, Butet. 2006. Sokola Rimba: Pengalaman Belajar Bersama Orang Rimba, Yogyakarta: INSIST Press.

Mardalis. 2002. Metode Penelitian Suatu Pendekatan Proposal, Bumi Aksara, Jakarta.

Marvin B. Sussmann (ed.). 1979. Marriage and Family, Collected Essays Series. New York: Hayworth Press.

Mertokusumo, Sudikno. 1993. Bab-bab Tentang Penemuan Hukum, Yoyakarta: Citra Aditya Bakti.

Muhammad, Bushar. 2000. Pokok-Pokok Hukum Adat, Pradnya Paramita, Jakarta.

Mulyana, Agung. 2006. "Perlindungan Hak -Hak Masyarakat Adat dalam Rangka Pembinaan Persatuan dan Kesatuan Bangsa", makalah disampaikan padaMusyawarah Lembaga Adat Rumpun Melayu se-Sumatera tanggal, April 14–17, 2006, di Riau.

M. Yunus Melalatoa. 1995. Ensiklopedi Suku Bangsa Di Indonesia. Jilid A-Z. Jakarta: Terbitan Departemen Pendidikan dan Kebudayaan.

Nasution, Khoiruddin. 2010. Hukum Keluarga (Perdata) Islam Indonesia. Yogyakarta: Academia Tazaffa.

Noel J. Coulson. 1967. History of Islamic Law. Edinburgh: University of Edinburgh Press.

Parsons, Talcott, 1951. The Social System: The Major Exposition of the Author & Conceptual Schema or the Analysis of Dynamics of the Social System. Canada: Collier Macmillan.

Parsudi Suparlan (Penyunting). 1993. Pembangunan yang Terpadu dan Berkesinambungan: Keterpaduan Pemanfaatan Sumber-Sumber dan Potensi Masyarakat Untuk Peningkatan Dan Pengembangan Pembangunan Masyarakat Pedesaan Yang Berkesinambungan.Jakarta: Terbitan Balitbangsos Depsos RI.

Prodjodikoro, Wirjono. 1976. Hukum Warisan di Indonesia, Sumur, Bandung.

Purba, Rehngena. 2000. Sikap Mahkamah Agung Terhadap Kedudukan Duda dan Janda Dalam Hukum Adat, Kanun No. 35 Edisi April.

Radcliffe-Brown. 1980. Struktur dan Fungsi Dalam Masyarakat Primitif. Malaysia Kuala Lumpur: Dewan Bahasa Dan Pustaka Kementerian Pelajaran.

Rostow, W.W. 1962. Process of Economic Growth. New York: W. W. Norton.

Simarmata,Rikarda. 2006. Pengakuan Hukum Terhadap Masyarakat Adat Di Indonesia.

Suwardi. 2006. Pemetaan Adat Masyarakat Melayu Riau Kabupten/Kota se- Provinsi Riau, Pekanbaru: Unri Press.

Suwarto. 2006. Mengangkat Keberadaan Hak-Hak Tradisional Masyarakat Adat Rumpun Melayu Se-Sumatera, Pekanbaru: Unri Press.

Kartodirdjo, Sartono. 1983. Metodologi Penelitian Masyarakat, Jakarta Gramedia.

Sagala. S. 1986. Majalah Budaya Batak Dan Pariwisata No. 8, Yayasan Budaya Batak Medan.

Santoso, Ananda. 1992. Kamus lengkap Bahasa Indonesia, Alumni, Surabaya.

Santrock, J.W. 2002. Life-Span Development: Perkembangan Masa Hidup. Jakarta: Erlangga.

Saragih Djaren, Samosir Djisman, Sembining Djaya, 1980. Hukum Perkawinan Adat Batak Khususnya Simalungun, Toba, Karo, dan Undang-undang Tentang Perkawinan (UU No. 1/1974), Tarsito, Bandung.

Sarwat, Ahmad. 2010. Fiqih Mawaris. Jakarta: Du Center.

Sharma, Arvind (ed.). 2002. Women in the World Religions. Terj. Syafa'atun Almirzanah. Perempuan dalam Agama-agama Dunia. Yogyakarta: Direktorat Perguruan Tinggi Islam bekerja sama dengan McGill University-CIDA Project.

Shihab, M. Quraish. Tafsir al- Misbah: Kesan, Pesan dan Keserasian al-Qur'an. Jakarta: Lentera Hati, 2002.

Sitepu Sempa. 1996. Sitepu Bujur, Sitepu A.G., Pilar Budaya Karo, Medan.

Sitohang Jailani dan Sibarani Sadar, 1988, Pokok-Pokok Adat Batak Toba (Tata Cara Perkawinan di Toba), March 26, Jakarta.

Soekanto, Soerjono. 1981. Meninjau Hukum Adat Indonesia, CV. Rajawali, Jakarta.

_____. 1986. Pengantar Penelitian Hukum, UI Press, Jakarta, 1986.

_____. 2001. dan Sri Mamuji, Penelitian Hukum Normatif Suatu Tinjauan Singkat, Penerbit Raja Grafindo Persada, Jakarta.

Soemadiningrat. 2002. Otje Salman, Rekonseptualisasi Hukum Adat Kontemporer Telaah Kritis terhadap Hukum Adat sebagai Hukum yang Hidup dalam Masyarakat, Alumni, Bandung.

Soemarman. 2003. Anto. Hukum Adat Perspektif Sekarang dan Mendatan, Adi Cita, Yogyakarta.

Soemitro Ronny Hanitijo. 1985. Metodologi Penelitian Hukum, Ghalia Indonesia, Jakarta.

Soepomo. 1967. Bab-Bab Tentang Hukum Adat, Penerbitan Universitas, Jakarta.

Sudiyat. Iman. 1981. Hukum Adat Sketsa Asas, Liberty, Yogyakarta, 1981.

Sutopo. H. B. 1998. Metodologi Penelitian Hukum Kualitatif Bagian II, Surakarta: UNS Press.

Syarifuddin, Amir. 1990. Pembaharuan Pemikiran dalam Hukum Islam. Padang: Angkasa Raya, h.17.

Tambunan E. H. 1982. Sekelumit Mengenal Masyarakat Batak Toba Dan Kebudayaannya Sebagai Sarana Pembangunan, Tarsito, Bandung.

Whaling, Frank (ed.). 1997. Contemporary Approach to the Study Religion. Melbourne: University of California Press.

Widen, Kumpiady. 2005. Impacts of Peatland and Forest Fires on Local Communities: Ecological, Health, Economics, and Cultural Perspectives. Dalam Tropical Petalands, International Journal For.

Widjaja, A. W. (ed.). 1986. Manusia Indonesia: Individu, Keluarga dan Masyarakat Jakarta: Penerbit Akademika Pressindo.

Wignjosoebroto, Soetandyo. 1994. Dari Hukum Kolonial ke Hukum Nasional Dinamika Sosial Politik Dalam Perkembangan Hukum di Indonesia, Jakarta: Raja Grapindo Persada.

Vergouwen. J. C. 2004. Masyarakat dan Hukum Adat Batak Toba, LkiS. Yogyakarta.

b. Laws

UU No. 39/1999 Tentang HAM.

UU No. 7/1984 tentang Pengesahan Konvensi Mengenai Penghapusan Segala Bentuk Diskriminasi Terhadap Wanita.

UU No. 23/2004 Tentang Kekerasan Dalam Rumah Tangga.

c. Articles

Badan Pembinaan Hukum Nasional (BPHN), Seminar Hukum Adat dan.

Pembinaan Hukum Nasional. 1976.Biro Pemberdayaan Perempuan Serdapropsu, Seminar Kebijakan Pemerintah.

Dalam Pembangunan Pemberdayaan Perempuan Pada Kegiatan Sosialisasi Gender, 2002.

Mahkamah Agung. 1979. Penelitian Hukum Adat Tentang Warisan Di Pengadilan Tinggi Medan, Mahkamah Agung Proyek Penelitian Hukum Adat, Jakarta.

d. Journals

Awaluddin, Yusuf Iwan. 2004. "Peningkatan Kepekaan Gender dalam Jurnalis" Jurnal Ilmu Sosial dan Politik. FISIPOL UGM, Vol. 7, No. 3, Maret.

Ditransliterasi dari.1992. Al-Allamah al-Bannany, Hasyiyah al-Bannany 'ala Syarh al-Mahally 'ala Matn Jam' al-Jawami. 'Beirut: Dar al-Fikr, jilid I, h. 25.

Waardenburgh, Jacques. 1973."Classical Approaches to the Study of Religion: Aim, Methode, and Theories of Research." The Hague: Mountor & Co., NV. 2 vols., Tahun.

e. Internet

http://intanghina.wordpress.com/2009/02/23/perkawinan-campuran-perlindungan-hukum-perempuan-wni-yang-melangsunkan-perkawinan-campuran/(accessed September 25, 2016).

http://bangdenjambi.wordpress.com/perkawinan-beda-agama-dan-hak-asasi-manusia-di-indonesia/ (accessed September 2016).

http://sonny-tobelo.blogspot.com/2009/02/fenomena-hukum-perkawinan-beda-agama.html (accessed September 25, 2016).

Chyntia Selvi Anggraeni at: http://www.scribd.com/doc/172648526/Prismatik-Society-Riggs

Emerging Trends in Psychology, Law, Communication Studies, Culture, Religion, and Literature in the Global Digital Revolution – Setiawan & Rahmawati (eds)
© 2020 Taylor & Francis Group, London, ISBN 978-1-03-224216-3

Duality of structure in the Family Welfare and Empowerment organization (PKK)

Ida Wiendijarti, Hermin Indah Wahyuni & Roso Witjaksono
Gadjah Mada University, Yogyakarta, Indonesia

ABSTRACT: This research is motivated that women have great potential in poverty allevi-ation activities through community and group empowerment. One of the organizations that is expected to become a potential container for development is Family Welfare and Empower-ment (PKK). The demand for PKK's flexibility to become a development agent and change agent is stronger, because PKK is one of the organizations working at the civil society level in improving family welfare and more effective because of the cadres reaching up to the village level. In today's democracy, it is interesting to examine how the PKK works, as an organiza-tion formed by the government from the central to the village level, capable to act as an ideal female empowerment agency. The use of the structural theory from Giddens is used to look at the relationships of structures and actors within the PKK's institutions, in its ability to respond to new challenges that come with economic reform and liberalization, in which increasingly diverse, universal and gender-responsive issues are present very quickly, so there needs to be a substantive renewal in the work program and organizational management in the future. Data were collected using a qualitative approach, with in-depth interviews and partici-patory observation, in the PKK of Sleman District, the results obtained in the form of analysis of the relation of structure and actors in PKK organization, which will be the reference for PKK development in the future, so PKK can act as empowerment agent adaptive women.

Keywords: structure, actors, organization, empowerment

1   INTRODUCTION

The Family Welfare and Empowerment Organization (PKK) is a women's organization that works with the government to improve social welfare, starting from the family level, to the provincial level and even at the central level. In the New Order era, the PKK was used as a political tool and as a means to oversee the ongoing ideology of the government (Suryakusuma, 2011), and provide significant support to the regime in power. Various social groups, including laborers and farmers. Women, especially those in rural areas, were co-opted through the PKK (Reeve, 1985).

Through the Decree of the Minister of Home Affairs No. SUS/3/6/12, on December 27, 1972, the government changed the name of the PKK from Family Welfare Education (Family Welfare Education) to Family Welfare Development. This change of name is considered cap-able of representing the state hegemonization of the women's movement in Indonesia (Han-dayani *et al.*, 2008), as an effort to ensure the loyalty of corporate state organizations.

The existence of political, economic, and social changes in Indonesia and Indonesia's commit-ment to achieve the Millennium Development Goals have an impact on the development and operational implementation of the PKK. Demands for PKK flexibility to become agents of development and agents of change are getting stronger. PKK's role in the community is one of the icons of civil society's movement in improving family welfare and its effectiveness is felt by the presence of cadres reaching the village/hamlet level. Until now the PKK is still the only women's movement in Indonesia that is able to break through to the village and hamlet levels, because the PKK is very close to the government bureaucracy (Soetjipto and Adelina, 2013).

As a movement that is expected to grow from the bottom, the management of the PKK through existing regulations, the driving team is a structure that is attached to the government bureaucracy. The relationship between structure and agency in PKK institutions is one of the interesting things to be developed. Structures (Giddens, 1984) are conceptualized as rules (resources) and resources that show the social practice of menstruation a long time and space, the structure of meaning will only be realized with relation to resources. Rules are defined as social agreements on how to act, while those resources discuss capabilities to make things happen, Structures are able to overcome space and time requiring structures not to exist in space and time, but these social practices that exist and continue in space and time (Herry-Priyono and Giddens, 2002).

There are three structural dimensions in the social system, namely: significance, dominance, and legitimacy (Giddens, 2010). Based on strengthening these three structural elements, a process of empowering PKK members by the government is aimed at the independence of the community and women, especially ins various productive activities. The researcher used the concept of routinization and awareness in structuration theory, which included unconscious motivation, discursive awareness, and practical awareness. The level of awareness that Giddens distinguishes into three determines the pattern of action that agents use in interacting. First, the agent is considered to have knowledge of most of his actions and this knowledge is expressed as practical awareness (Kaspersen and Sampson, 2000). Practical awareness refers to a group of practical knowledge that cannot always be described (Herry-Priyono and Giddens, 2002). Practical awareness contains knowledge that has been assumed and taken for granted (taken for granted knowledge). Through a group of practical knowledge, agents know how to live daily without having to constantly question what will happen or what should be done (Cohen, 1989). Personal and social life routines are formed through the performance of this practical awareness group.

The researcher chose Sleman Regency, to see the dynamics related to the dynamics of the PKK organization with the structure inherent in the rules of the PKK Movement. Organizational dynamics are seen through the concept of awareness and routinization made in the organization so far in the implementation of the PKK movement routine, both in routine and incidental activities.

2   LITERATURE REVIEW

Structural theory focused on three main concepts regarding "structure", "system", and "structural nature" (Giddens, 2010), especially in relationships between agents (actors, actors) and structures. Through his structuration theory Giddens emphasizes the study of "ongoing social practices" as he stated, that "the basic domain of the study of the social sciences, according to structuration theory, is not the experience of individual actors, nor the existence of any form of social totality, but the practice arranged in throughout time and space (Goodman and Ritzer, 2008), (Giddens, 1984), (Lamsal, 2012). Structuration sees the importance of social practices, both in action and in the structure of people's lives. Structuring refers to "a way in which social structures are produced, reproduced, and changed in and through practice". Structural duality lies in the process by which structure is of an outcome and at the same time becomes a medium (medium) of social practice, meaning duality of structure lies in the fact that a structure which is the principle of social practices in various places and times is a result of continuous views of various social practices carried out by agents, and at the same time, the structure is also a medium for the agent's social practice to take place. (Herry-Priyono and Giddens, 2002) In this case the structure and agents carry out interactions that influence each other. Through duality of structure, relationships between structures and agents can be seen, where agents with the range of knowledge they have can make the structure a reference in acting, and change and reproduce structures through routine social practices. The structure is actively produced, reproduced and changed by agents who are seen as actors who have the ability.

Structural theory emphasizes the interrelationships of structures and agents in a relation of duality. Structure and agents are interrelated without being separated in human social practice (DeSanctis and Poole, 1994). Gidden emphasizes that agents are people who are involved in continuous flow of action, meaning that agents are actors in social practice, where agents are seen both as individuals and as groups (Herry-Priyono and Giddens, 2002). Structuring in a group is described as a process by which systems are produced and reproduced through the use of rules and resources by members (Poole, Seibold and McPhee, 1996). Structuring is the production and reproduction of social systems through the use of rules and resources by members in interaction. Interaction in the theory is an action based on free will. While rules are propositions that make decisions worth or show how something should be done. While resources are materials, property, and characters that can be used to influence or control the actions of groups or members. Production occurs when group members use rules and resources in interactions, while reproduction occurs when an action strengthening feature of the system is in place.

3 METHOD

Related to the reality of the women's empowerment movement carried out through the PKK, as well as government dynamics and policies in the context of women's empowerment, this study uses a constructivist/interpretive paradigm. The research subjects were PKK organizations from the district level to the villages in Sleman Regency.

4 RESULTS AND DISCUSSION

Family Welfare Empowerment (PKK) is a women's movement in Indonesia that is able to reach up to the village and hamlet levels, even hamlets from all regions in Indonesia. This can happen because the PKK is very close to the government bureaucracy which is their own husbands. Internal challenges within the organization include: leadership and the potential of human resources, which cannot work optimally; the governance of the PKK movement and the program, still depends on the leadership that is on duty and the conditions of each region. These results from the structure and agent relations developed PKK programs which became the realization of the PKK movement, where the existing structure was a manifestation as well as a communication product within the PKK organization. Each PKK program must be preceded by a proposal submitted with each SKPD to obtain funding, even though there are also PKK that receive Village Fund Alocation (ADD).

Meanwhile, the external challenges faced by the PKK, namely: post-Reform political change; and relations between institutions and government and private institutions in an effort to harmonize with PKK programs for community empowerment. The results showed that PKK institutions were collective actions of women who became PKK cadres with government bureaucracy from hamlet to central level, where they moved on various strategic choices which were a function of the dialectic between structure and agency. Reciprocal relationships that occur between agents and structures in social practice are repeated and patterned in space and time. Repeated social practices of individual agents who reproduce these structures through PKK programs.

The struggle of PKK cadres as the spearhead of the women's movement in the village, it is appropriate to get an award. There are still many PKK cadres in remote villages who selflessly struggle to overcome the problems faced by village women. Without understanding the concept of gender or the interests of women, they continue to move to carry out their responsibilities and caring for the community.

The implementation of the Village Law must be supported by implementing regulations which include government regulations, ministerial regulations, regional regulations and village regulations. During this time, the women's program only focused on the Ten Principal PKK Programs and revolved around domestic issues. The ideology took place continuously, tiered from the wife of the village head to the residents. PKK activists and PKK cadres find it difficult to get out of old ways and progressively think.

Gender issue talks must be attached to other issues, such as: poverty, education, health, and sustainable development. Building gender responsive villages by using indicators of the Human Development Index (HDI), Gender Development Index (GDI), and Gender Empowerment Measure (GEM), whose parameters are innovatively related to AKI, AKB, economy, family planning, trafficking, domestic violence, education, eradication of endemic infectious diseases, environmental destruction, and representation of women in the policy making positions.

In Indonesia, the state's patriarchal ideology severely limits women's role only to the concept of reproduction in the household. This ideology, consciously or unconsciously, likes or dislikes, seems clear or vague, penetrates all joints of life, and forms a layered discrimination system that is difficult to conquer. This ideology is even implanted by the state in layers from the central, village and even hamlet levels. The manifestation can be seen from the Ten PKK Principal Programs. The project of nationalism, development and democracy continues to develop and run through the country's political power. Interested state reaches its goal, forming a patriarchal ideology that is parallel to that goal. Autonomy and independence are hardly visible between women's organizations and women's activist organizations.

In Indonesia, the state's patriarchal ideology severely limits women's role only to the concept of reproduction in the household. This ideology, consciously or unconsciously, likes or dislikes, seems clear or vague, penetrates all joints of life, and forms a layered discrimination system that is difficult to conquer. This ideology evens PKK was used as a mobilization tool for development purposes by encouraging the principle of volunteerism. The thing that makes PKK cadres work in villages across borders, Ethnicities, religions, and classes is the conception of motherhood that prioritizes the interests of children, husbands, extended families, and nearby communities as a binding identity. But when that identity is linked to women's issues which are the concern of village women, the answer is no single. Implications at the rural level: PKK has difficulty responding to new challenges that are present along with economic reform and liberalization. Issues that are increasingly diverse, universal, and gender responsive are very fast. The village PKK has not experienced substantive renewal at work programs, organizational management, leadership models, and lagging behind in adapting to new environmental changes and new issues that are very rapidly developing.

Organizing activities is understood as a new structure that is formed through dialectic structure and agency which has an impact on changing old structures to new, more adaptive structures (which one party restricts and the other accustoms to) and agency (which has transformative capacity) to modify the structure with the final goal of its production new patterned actions that are considered more relevant to the relationship between goals and means, both those that have been available or can be provided or those found. This results in the actions of agents in modifying and transforming structures can produce a variety of new patterned patterns that have positive or negative aspects, or both at the same time for the existence and continuity of the organization.

Organizing institutions is the result of agents' collective action on a variety of strategic choices that are a function of the dialectic between agency and agency, and between constraints and opportunities. What is understood by informants as 'acrobatic actions', actually reflects the truth of the Giddens proposition, namely agents are knowledgeable individuals. In addition, this also shows that agents have the power to change barriers to possibilities.

As a social practice, the planning and implementation of PKK programs through the PKK movement is carried out by PKK cadres to overcome budget constraints and lack of adequate knowledge in handling cases such as trafficking, domestic violence and other women's welfare issues, as well as outside who in 10 PKK Principal Programs, representing:

1. Authoritative sources, namely the role of the PKK driving team, provide normative authority that can be used as a basis for transformative capacities that make them change the character of allocative sources (such as limited budgets, personnel and facilities) that initially exclude the possibility or limit action to the contrary.

2. Each agent has a different transformative capacity for each context of sociality and individuality. Position, position and role are important authoritative sources for generating power. However, rules can make authoritative and allocative sources relative to actions because only through the decisions of agents is an action possible.
3. Transformative capacity is something that is inherent in agency which causes a relatively unlimited possibility to produce action.

The complexity of duality is accompanied by constraints and opportunities or between structure and agency lies in the transformative capacity inherent in genes whose consequences cannot be predicted beforehand. The agent's transformative capacity in converting constraints into opportunities, or at different times, taking advantage of opportunities to produce a breakthrough has very important implications for structural modifications.

Agents will continue to find breakthroughs as long as the rules and other authoritative sources benefit the possibility of action. The role as a PKK cadre with its various authorities based on the basic rules of the PKK, allows agents to transform their formal power into a relatively almost unlimited capacity to produce creative actions to transform obstacles into opportunities.

5 CONCLUSION

Substantial revitalization and renewal of PKK seems necessary, especially ideology and paradigms related to handling gender issues, changes in internal management, internal democracy in organization and leadership, and fundamental budget political reform, so that the PKK can continue to be relevant in the future. The PKK has enormous potential, if it is able to transform and change the face of the village from underdevelopment, poverty, and underdevelopment, to become a prosperous, empowered, and faired village, so that it can become a strong foundation for development. The author suggests that reforms in PKK institutions must start from work programs, manage organizations more professionally with the selection of competent leaders, and be able to adapt to increasingly rapid environmental changes in the global era.

REFERENCES

Cohen, I. J. (1989) *Structuration theory: Anthony Giddens and the constitution of social life*. Macmillan International Higher Education.
DeSanctis, G. and Poole, M. S. (1994) 'Capturing the complexity in advanced technology use: Adaptive structuration theory', *Organization science*. INFORMS, 5(2), pp. 121–147.
Giddens, A. (1984) *The constitution of society: Outline of the theory of structuration*. Univ of California Press.
Giddens, A. (2010) 'Teori Strukturasi: Dasar-dasar Pembentukan Struktur Sosial Masyarakat', *Yogyakarta*: *Pustaka Pelajar*.
Goodman, D. J. and Ritzer, G. (2008) 'Teori Sosiologi', *Jakarta*: *Kreasi Wacana*.
Handayani, T. *et al.* (2008) 'Pemberdayaan dan Kesejahteraan Keluarga (PKK) di kota Malang: dalam Perspektif kajian Budaya', *E-Journal of Cultural Studies*, 2(1).
Herry-Priyono, B. and Giddens, A. (2002) 'Suatu Pengantar'. Jakarta: Gramedia.
Kaspersen, L. B. and Sampson, S. (2000) *Anthony Giddens: An introduction to a social theorist*. Blackwell publishers Oxford.
Lamsal, M. (2012) 'The structuration approach of Anthony Giddens', *Himalayan Journal of Sociology and Anthropology*, 5, pp. 111–122.
Poole, M. S., Seibold, D. R. and McPhee, R. D. (1996) 'The structuration of group decisions. RW Hirokawa, MS Poole, eds. Communication and Group Decisionmaking'. Sage Publications, Thousand Oaks, CA.
Reeve, D. (1985) *Golkar of Indonesia: An alternative to the party system*. Oxford University Press, USA.
Soetjipto, A. W. and Adelina, S. (2013) *Suara dari desa: menuju revitalisasi PKK*. Marjin Kiri.

Emerging Trends in Psychology, Law, Communication Studies, Culture, Religion, and Literature in the Global Digital Revolution – Setiawan & Rahmawati (eds)
© 2020 Taylor & Francis Group, London, ISBN 978-1-03-224216-3

Blonde hair as symbol of Patrick Bateman's displacement in *American Psycho*: Semiotic approach

Dini Anggraheni
Universitas Semarang, Semarang, Indonesia

ABSTRACT: The research aims to reveal the meaning of "blonde hair" symbolism, and its relationship with Patrick Bateman's defense mechanism in the film *American Psycho*, directed by Mary Harron in 2000. The main source of this research is a DVD film downloaded from www.thepiratesbay.se using Torrent application software. The reason it becomes the main source of this research, because it has a complete scene without editing or cutting the scene.

This research uses quantitative methods. The data analysis of this research uses structuralism theory from A.J Greimas, whose actantial scheme uses the narrative to determine the semiotic aspects of a symbol along with Patrick Bateman's victims in the film and what drives Patrick Bateman to kill (or to fantasize about killing).

In order to strengthen the results of this research, psychology analysis from Freud is used to explain the defense mechanism of Patrick Bateman. The hypothesis of this analysis is about displacement, one of the defense mechanism types released by Patrick's love, hate, suppressed, murder instinct, killing satisfaction, etc.

Keywords: blonde hair, *American Psycho*, semiotics, psychoanalysis, displacement

1 INTRODUCTION

In the mid-19th century, what was referred to as "fine arts" was a general translation of literature and music (Monaco, 1981) so that film as a work of art is a literary work. James Monaco stated that film is not a language but is like language, so some ways of learning languages are possible when using film as a language teaching material.

A film is a combination of several other elements of art. Film utilizes elements of technology, environment, image, dramatic, narrative and music as media representations (Monaco, 1981). The study of film must take into account the various aspects that accompany the narrative. Bordwell, Thompson and Smith (1997) state that films are discussed in terms of narration and style, which will support a total understanding of the film (Monaco, 1981). *American Psycho* is a realistic reflection of the mental problems experienced by Patrick Bateman, who is described as a successful young executive. In some movie scenes, blonde hair is believed to be symbolic of the psychiatric disorders of the main character, Patrick Bateman.

2 LITERATURE REVIEW

David Robinson's research entitled 'The unattainable narrative: identity, consumerism and the slasher film in Mary Harron's *American Psycho*' was a reference for developing this analysis of the film. Robinson's research is in the form of 26 articles published in the journal *Cine Action 68* (Winter 2006), in which he describes various narratives of novels that cannot be revealed in films. However, Mary Harron has another power that is almost the same as Alfred

Hitchcock in his film, which is the ability to make movie lovers find something more "murderous" than when reading the novel.

Robinson (2006) also explains that psychology can reveal what happened to Patrick, especially psychosis. To discover aspects of Patrick's psychology, we use film according to Miarso (2004). The most sophisticated media can convey five types of information: images, lines, symbols, sounds, and movements. In this film directed by Mary Harron, we used the symbolism that appears and analyze it using semiotic Greimas.

Greimas (1966) through Teeuw (1988) is a French researcher with structural theory (Lithuanian semiotician structuralist). During the 1960s, Greimas (1966) proposed the actantial model, a tool that can theoretically be used to evaluate any real or thematized action, but particularly those depicted in literary texts or images. In the actantial model, an action may be split into six components, called actants. Actantial analysis subsists of assigning each element of the action being described to the various actantial classes.

Hébert (2019) stated that axis of desire is (1) subject and (2) object. The subject is what is directed toward an object. The rapport enacted between the subject and the object is called a junction, and can be further classified as a conjunction or a disjunction. The axis of power is (3) helper and (4) opponent. The helper assists in achieving the desired junction between the subject and object; the opponent hinders this. The axis of transmission (the axis of knowledge, according to Greimas) is (5) sender and (6) receiver. The sender is the element requesting the establishment of the junction between subject and object. The receiver is the element for which the quest is being undertaken. To simplify, let us interpret the receiver as that which benefits from achieving the junction between subject and object. Sender elements are often also receiver elements.

Technically speaking, we need to distinguish the actantial model as a conceptual network from its visual representation (*American Psycho* film scene). The conceptual network is generally depicted as a diagram (Figure 1).

Self-defense mechanisms that shift sexual or aggressive impulses to more acceptable or less threatening targets; direct emotions to safer objects; emotional separation from real objects and redirection of intense transfer emotions to the weaker, this is done to directly avoid things that are scary or threatening for him/her (Semiun, 2006). In other words, displacement is directing energy to other objects or people if the original object or real person cannot be reached.

Based on the description, we hypothesized that:

H1: Patrick kills all his victims related to their blonde hair.
H2: Blonde hair is a symbol of Patrick Bateman's displacement.
H3: The displacement is about Patrick's fiancé, Evelyn.

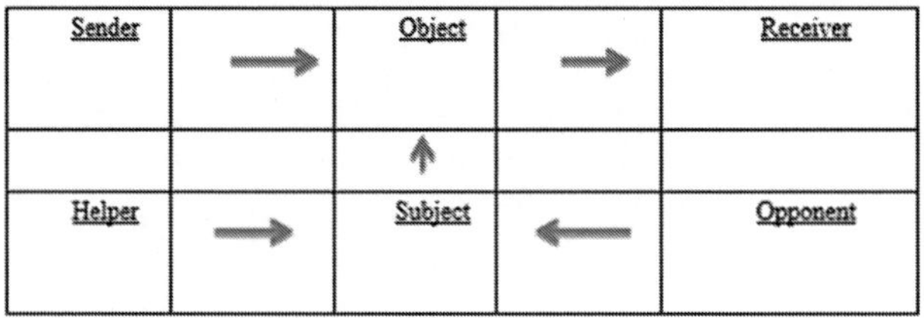

Sender	→	Object	→	Receiver
		↑		
Helper	→	Subject	←	Opponent

Figure 1.   The conceptual network.

3   METHOD

This research uses quantitative methods. The data of this study is the symbolism from Greimas' actantial scheme and it analyzed with psychoanalysis theory to show the sympthoms of displacement.

4 RESULTS & DISCUSSIONS

The analysis process uses Greimas' actantial scheme on five of Patrick Bateman's victims. The first victim is Daisy, a blonde model: Patrick met Daisy in the Night Club. Blonde (sender) is something that makes Patrick interested in Daisy (object) a beautiful model. With his good-looking face (helper), Patrick invites Daisy to go with him. Because of the crowd at the night club (opponent), Patrick had a reason to invite Daisy to go by taxi to his house. In a later scene, Patrick is seen playing with similarly colored blond hair that is kept in his pocket and the head of a blonde girl is wrapped in plastic in the refrigerator. This could be a sign that Patrick had killed Daisy and took her hair.

Explanation of the second victim analysis using Greimas' actantial scheme. Jean is a secretary in his office. Jean who has blonde hair (sender) makes Patrick (subject) approach and invite her to his apartment. Beginning with small talk, Jean felt comfortable in Patrick's apartment. When chatting, Patrick uses a nail gun (helper) to kill Jean (receiver). Suddenly the phone rang and Patrick heard Evelyn's voice:

> Patrick I know you're there. Pick up the phone, you bad boy. What are you up to tonight? It's me. Don't try to hide. I hope you're not out with some little number you picked up because you're Mr. Bateman. My boy next door. Anyway, you never called me and you said you would and I'll leave a message for Jean about this too to remind you, but we're having dinner with Melania.

After Evelyn's short message, Patrick felt anxious, abandoned his plan to kill and asked Jean to leave. Thus, Evelyn has a power to control Patrick's emotions.

Explanation of the third victim analysis using Greimas' actantial scheme. Sabrina is a prostitute. Pat-rick called someone to find a prostitute:

> I'd like a girl, early twenties, blonde, who does couples. Couples. Fifty-five west Eighty-First, the American Gardens Building, apartment 7C. and I really can't stress blonde enough. Blonde.

Sabrina (object) is a blonde prostitute (sender) even though a dark blonde type was asked by Patrick (subject) to become his sex partner. After aberrant sexual intercourse, Patrick (receiver) hurts Sabrina using sharp objects (helper) until her service time was run out (opponent).

Explanation of the fourth victim analysis using Greimas' actantial scheme. Elizabeth is Pat-rick's co-worker. Elizabeth (object) who has dark blond hair (sender) makes Patrick want to vent his sexual desire (receiver). With the help of stimulants (helper), Patrick makes Elizabeth desire sexual relations. After sexual intercourse, Patrick bit Elizabeth's ears until she was seriously injured. Elizabeth shouted (opponent) and that made Patrick furious.

Explanation of the fifth victim analysis using Greimas' actantial scheme. Christie is a prostitute. Christie (object) is a prostitute who has been hired twice by Patrick (subject) and has similar hair to Evelyn (sender). She refused Patrick's invitation, but changed her mind after Patrick offered a lot of money (helper) and said he would be nice this time. After Christy saw Patrick's sexual deviations, he resisted by running out of the room (opponent). Patrick then killed Christie using a saw machine and keep her body in the closet (receiver).

According to Greimas' actantial scheme, the sender is similar in the five victims. Blonde hair is the real symbol that becomes a reason for Patrick's displacement.

5 CONCLUSION

Based on the analysis carried out in the film *American Psycho*, it can be concluded that there are several phenomena that cumulatively lead to the main hypothesis. The first phenomenon is the psychological condition of Patrick Bateman, who is identified as using displacement as a defense mechanism, which is the effect of psychological repression by his lover named Evelyn.

> *Evelyn Williams, Patrick Bateman's fiancée, is making notes with gold cross pen. Evelyn is blonde, classically beautiful, expensively educated, and utterly pleased with herself. She usually addresses Patrick as if he were a small child.*

The second phenomenon is a form of displacement compensation, which is an obsession with revenge for psychological repression and is experienced by killing blond women. The third phenomenon is the meaning of the symbol of blonde hair, which is the strongest symbolic representation of Evelyn.

Psychologically it can be explained that the accumulation of psychological repression experienced by Patrick Bateman was caused by the behavior of his lover (Evelyn), which became a pressure in his subconscious. Several analyses that have been carried out. Our hypothesis is that blonde hair is a displacement index for Patrick Bateman to vent or channel all feelings of love, hatred, anger, and revenge. This index comes from Evelyn, who dominates and regulates Patrick's life. Narrative, semiotics, and psychoanalysis can become a unit that produce a unique analysis.

This study still has many limitations, such as not all hypotheses show the displacement symptoms. Therefore, it is recommended that further research also focuses on the symptoms of displacement that can be found in every scene and use the newest references.

REFERENCES

Bordwell, D., Thompson, K. and Smith, J. 1997. *Film art: An introduction.* McGraw-Hill New York.
Greimas, A. J. 1966. Semantique structurale Paris: Larousse. For a summary see Hawkes, T. *Structuralism and Semiotics*, 1977. Methuen New Accents, London, pp. 87–95.
Hébert, L. 2019. An Introduction to Applied Semiotics: Tools for Text and Image Analysis. Routledge, London.
Monaco, J. 1981. *How to read Film.* New York: Oxford University Press.
Robinson, D. 2006. The Unattainable Narrative: Identity Consumerism and the Slasher Film in Mary Harron's 'American Psycho', *CineAction*. CineAction Collective, pp. 26–35.
Semiun, Y. 2006. Teori kepribadian dan terapi psikoanalitik freud, *Yogyakarta: Kanisius*, pp. 87–89.
Teeuw, A. 1988. Sastra dan Ilmu Sastra, Sebuah Pengantar Teori. *Jakarta Pustaka Jaya.*

Emerging Trends in Psychology, Law, Communication Studies, Culture, Religion, and Literature in the Global Digital Revolution – Setiawan & Rahmawati (eds)
© 2020 Taylor & Francis Group, London, ISBN 978-1-03-224216-3

Authority of religious court in settlement of the Shariah banking dissolution

Dhian Indah Astanti, B.Rini Heryanti & Subaidah Ratna Juita
Universitas Semarang, Semarang, Indonesia

ABSTRACT: Islamic banking principles are a part of Islamic education that relate to the economy. Several principles of Islamic economics are the usury restriction in any form and using any system, among others, in the form of profit-sharing principles. With the standard of benefit sharing, Islamic banks are able to establish a healthy and fair investment atmosphere where all the parties are able to share emerging benefits and potential risks, in order to create a balanced position between the bank and its customers. Given the development of Islamic banks so far, Sharia principles, which are the main foundation of Islamic banks in carrying out their duties, cannot be implemented and enforced optimally, especially in the event of a dispute between parties, Islamic banks, and their customers. The purpose of this research is to determine and understand the authority of the Religious Courts in finding a solution for the Sharia banking conflict and the principles of handling Sharia banking dispute settlements. This research is a sociological juridical legal research. The issuance of Law Number 3 of 2006 concerning Amendments to Law Number 7 of 1989 concerning Religious Courts since 30 March 2006, has provided a legal umbrella for the implementation of Sharia economics in Indonesia. Disputes in Sharia banking are the authority of the religious court environment, and dispute resolutions related to Sharia banking economic activities are completed in two ways: litigation and non-litigation. In addition, the issuance of Law Number 21 of 2008 concerning Islamic Banking further reinforces the dispute resolution mechanism between the bank and the customer as stipulated in Article 55 paragraph (1), (2) and (3) that dispute resolution is carried out in accordance with the contents of the contract.

Keywords: dispute resolution, Syariah banking

1 INTRODUCTION

Islamic banking in Indonesia has experienced rapid and impressive developments. The average growth of Islamic banking assets has reached more than 30% per year. From the legal aspect, the development of Islamic banking is also significant and is marked by the birth of Law Number 21 of 2008 concerning Islamic Banking.

Like conventional banks, Islamic banks have a function as intermediary financial institutions, which carry out a mechanism for collecting and channeling funds in a balanced manner, in accordance with the applicable provisions (Mohamme, 2012).

Therefore, the development and growth of Islamic Banking requires support from four aspects. First, strengthening aspects of government regulation in supporting the growth rate of the Sharia Economy in Indonesia. Second, the development of practical aspects of Islamic business and financial institutions. Third, the development of Islamic economic science through research, both individually and institutionally, such as the development of the Islamic Economics College and Islamic Economics Higher Education. Fourth, the acceleration of the growth of Islamic Economics institutions in Indonesia.

The enactment of Law Number 21 of 2008 concerning Sharia Banking, recognizes the existence of Islamic banks in Indonesia that carry out the functions of financial intermediary institutions in accordance with Sharia principles as an operational foundation. Sharia banks, according to Article 1 paragraph 7 of Act Number 21 of 2008, carry out business activities based on Sharia principles and according to their types and consist of Sharia Commercial Banks (BUS) and Sharia People Financing Banks (BPRS).

The principle of Islamic banking is part of Islamic teaching relating to the economy. One of the principles in Islamic economics is the prohibition of usury in its various forms, and using the system, among others, in the form of profit-sharing principles. With the principle of profit sharing, Islamic banks can create a sound and fair investment climate because all parties can share both the benefits and potential risks that arise so that they will create a balanced position between the bank and its customers.

To facilitate the discussion, this study aimed to identify the problems concerning the authority of the Religious Courts in resolving Sharia banking disputes and the principles of handling Sharia banking dispute resolution.

2 ANALYSIS AND DISCUSSION

2.1 *Legal aspects of implementing Sharia principles in financing contracts in Indonesian Islamic banking*

The term contract is often called a contract or agreement, which is a meeting of consent granted by one of the parties, with qabul legally given by another party according to Sharia law and creates legal consequences, namely the emergence of rights on the one hand and obligations on the other. From this understanding, establishing a contract is key in Islamic banking, because without a contract, the transaction is doubtful and may cause disputes at some point.

The financing contract implies one that is made by default, in which one party has prepared the standard requirements on the existing contract form and then offered it to the other party for approval with limited negotiation opportunities. The validity of the contract is determined by whether the clauses stated in the Sharia contract conflict with Islamic principles or not (Alamsyah, 2013).

Assessed from the aspect of private law, the banking system based on Sharia by applying the principle of profit-sharing in offering finance to customers, either through fundraising or channeling of funds, is the legal relationship between the bank and the customer. This is a contractual agreement or fund contract (sahib al-mal) with fund management investors (mudharib) who work together to conduct productive businesses and share profits fairly (mutual investment relationship). Therefore, when entering into various contracts Islam firmly and clearly encourages every legal subject consisting of individuals and individual's legal entities to be careful and always pay attention to harmony and the legal requirements of the contract as specified in Islamic law.

In Sharia business financing transactions, establishing a contract is key because without a contract, the transaction is doubtful and may cause disputes at some point The firm and clear incitement fully encourages citizens and especially the adherents to be careful and have a contract for every transaction they carry out.

Every financing transaction in an Islamic bank requires a contract. The contract happens at the beginning. In making financing contracts, many Islamic banks still refer to the format of credit agreements in conventional banks. However, adjustments were made in the articles so as not to conflict with Sharia principles.

Adjustments made are guided by applicable Islamic law, and then refer to the provisions of Indonesian Positive Law. Indonesian law, which is also noteworthy in the making of this agreement, includes the law on Sharia banking, limited liability company law, decrees of the Board of Directors of Bank Indonesia, Fatwas of the National Sharia Council (DSN), and so forth.

2.2 *Effectiveness of the application of Sharia principles in the compilation of Sharia business contracts*

The effectiveness of applying Sharia principles in Sharia business contracts is reflected in the capital owners and capital managers themselves. The financing effectiveness of the capital manager (customer) side is based on several parameters:

a. A financing procedure that shows the ability of prospective customers to understand it;
b. Funding requirements that show the ability for prospective financing customers to fulfill them, including the presence or absence of collateral;
c. Disbursement time or realization that shows the speed of Islamic banks to provide the proposed financing;
d. The location of banks for the convenience for customers to access the capital sources provided;
e. Impact of financing that shows the level of stability of financing.

If viewed from the side of capital management, the effectiveness of profit-sharing financing with the Mudharabah and Musyarakah principles can be measured through the distribution of funds. This is related to the extent to which the owner of the capital distributes financing with the Sharia system, meaning that the more funds channeled, the more effective the financing of the Sharia system.

The effectiveness of applying Sharia principles in business contracts in Islamic banking can also be measured by looking at the stability of financing procedures based on the following factors:

a. Number of customers who indicate that the financing system is acceptable and wide reaching;
b. A diversity of customer livelihoods that shows the flexibility of the financing procedures implemented;
c. The frequency of customer loans, as the level of frequency of customers taking financing;
d. Frequency of arrears, as the level of frequency of customers in delinquent payments in a loan process;
e. Financing services, the extent to which service levels are carried out starting from the filing of financing to the realization of financing.

The implementation of Murabahah financing in Islamic banks is not entirely based on Islamic principles that are adjusted to the applicable laws and regulations concerning Sharia, namely: Law Number 21 of 2008 concerning Sharia Banking, Bank Indonesia Regulation No. 6/24/PBI/2004 concerning Commercial Banks that carry out Business Activities based on Sharia Principles, and Fatwa National Sharia Council Number 04/DSN-MUI/IV/2000 concerning Murabahah. But it is also based on other positive laws. Therefore, the implementation of financing based on the Murabahah principle does not always run as determined and agreed to in the contract agreed upon by the parties. There are risks and concerns from the owners of capital on this murky financing: if the financing provided by banks to customers is not smooth it becomes problematic financing, which becomes a dispute between the customer and the bank. There is a need for banks to take special steps to save financing funds and in resolving disputed financing disputes between banks and customers, to prevent risks in Murabahah financing carried out by customers. The funds available at the bank not only come from the capital owner's funds, but also from the customers who deposit their money with the bank. Thus, it is appropriate for banks to maintain and account for the trust of these customers.

3 CONCLUSION

Law on Sharia Banking Number 21 of 2008 has given competence or authority to courts in the general court environment in resolving Sharia banking disputes and has reduced the absolute competence of religious courts. In Law Number 3 of 2006 it is very clear that religious courts have absolute competence in the field of Islamic economics, including concerning Islamic banks.

One of the provisions stipulated in Law No. 21 of 2008 is Article 55 Paragraph (1), which regulates the place for resolving sharia banking disputes. The article states "Sharia Banking Dispute Settlement is carried out by courts within the Religious Courts." However, the provisions of Paragraph (2) and Paragraph (3) of the article provide the opportunity for dispute resolution elsewhere. The conditions for the place of completion have been agreed upon by the parties in the contract.

With the presence of Sharia banking law, the competence of the court in handling problems in Sharia banking disputes is not only the authority of the religious court, asthe general court has the same authority to handle Sharia banking dispute cases.

There needs to be contributions from various parties to examine and review the Sharia banking law regarding the competency of litigation institutions in Sharia banking disputes. Thus, the mandate of Law Number 3 Year 2006, which gives full authority to religious courts in resolving economic disputes in Islam including Islamic banking, can be fully implemented.

REFERENCES

Alamsyah. 2013. *Judge of the Sengeti Religious Court, 'Exemption Clause in Sharia Standard Con-tracts'.* Available at: www.badilag.net (accessed: 21 February 2013).

Mohamme. 2012. *Sharia Bank System and Operating Procedure.* Yogyakarta: UII: Press.

Anshori, Abdul Ghofur. 2010. Sharia Banking Law. Bandung: Refika Aditama. Establishment of Sharia Banks Through Acquisitions and Conversions. Yogyakarta: UII Press.

Cak Basir. 2012. Sharia Banking Dispute Settlement in PA and Mahkamah Syariah. Jakarta: Kharisma Putra Utama.

Dewi Gemala. Legal aspects of Sharia Banking and Insurance in Indonesia. Jakarta: Prenada Media.

Muh Firdaus, et al. 2005. Sharia Supervision System and Mechanism. Jakarta: Renaissance.

Muhammad. 2002. Bank Syariah Management. Yogyakarta: UPP AMP YKPN.

Sutedi Adrian. 2014. Legal Aspects of the Financial Services Authority. Jakarta: Achieve Asa Sukses.

Sutiyoso Bambang. 2006. Business Dispute Resolution: Solutions and Anticipation for Business Enthusiasts in the Face of Current and Future Disputes. Yogyakarta: Citra Media.

Sudarsono Heri. Islamic Financial Banks and Institutions Descriptions and Illustrations, Cet. Yogyakarta: Ekoni-sia-FE UII.

Soekanto Soejono. 1986. Introduction to Legal Research. UI: Press.

Soekanto Soerjono and Sri Mahmudji. 1996. Normative Legal Research A Brief Overview. Jakarta: PT. Graf Grafindo Persada.

Soemitro, Ronny Hanitijo. 1990. Legal and Jurimetric Research Methodology. Jakarta: Ghalia Indonesia, Jakarta.

Syahdaeni, Sutan Remi. 1999. Islamic Banking and Its Position in the Indonesian Banking System. Jakarta: Main Li-brary of Graffiti.

Zuhri M. 1995. Riba in the Qur'an and Banking Issues: An Anticipatory View. Jakarta: Raja Grafindo Persada.

Emerging Trends in Psychology, Law, Communication Studies, Culture, Religion, and Literature in the Global Digital Revolution – Setiawan & Rahmawati (eds)
© 2020 Taylor & Francis Group, London, ISBN 978-1-03-224216-3

Implementation of well-designed interior spaces, with psychological concepts and solutions for private workspaces in Surabaya

Sriti Mayang Sari & Sherly de Yong
Department of Interior Design, Petra Christian University, Surabaya, Indonesia

ABSTRACT: A well-designed workspace can increase productivity and enable efficient use of space. Despite the well-designed space concept, there are also cultures and habits that indirectly affect the use of interior space functions and the atmosphere in the workspace. Therefore, this research aimed to identify the issues with implementing well-designed space for private workspaces in Surabaya City. The case study for this project is portfolios were designed by students Regina, Aileen and Leony from the Interior Design Department – Petra Christian University. The research method was qualitative research. The method used to collect primary data was direct observation, interviews, and literature studies. Analysis was carried out by comparing literature data with field data. This analysis created a solution for the implementation of well-designed interior space in the directors' room in Surabaya. The design will help the workers to improve their performance, for example adding plants could provide a sense of comfort and creativity.

Keywords: environment psychology, interior, well-designed space, workspace

1 INTRODUCTION

Humans try to fulfill their lives by working, which becomes one of their important activities. A workspace is a space used or required for one's work, in an external space or at home. Therefore, a workspace has an important role in accommodating human activities. Good interior workspace design can help workers improve productivity. Besides being able to increase productivity, good workspace arrangements can also provide other benefits, such as enabling the efficient use of spaces and facilitating communication between users so that coordination and supervision are easier. In addition to the effectiveness of work, good workspaces can affect user discipline (Van Meel, Martens and van Ree, 2010).

The interior design of effective workspaces requires careful consideration of many issues., from the sensory input aspects, organizational and national cultures, through the psychological needs for privacy (Augustin, Frankel and Coleman, 2009). Creating effective interior workspaces, will increase workers' satisfaction with their jobs, which is important for humanitarian and financial reasons. Based on Augustin, Frankel and Coleman (2009), when someone is satisfied with their workspace, they will be more happy with their work. Therefore, a well-designed workspace should include five psychological aspects: communicating, comforting, complying, challenging and continuing. Some studies have stated that the way architects respond to building space is completely different from the needs, culture and habit of users. Moore in Snyder and Anthony (1991) stated that architects are usually more concerned with the physical and surrounding environments in building space (Snyder and Anthony, 1991). Andi Siswanto in Sari stated that the most important thing is development which must focus on the values and socio-culture of the local community (Ishananto, 2010; Sari, 2018). Therefore, beside the five psychological aspects from Sally Augustin regarding the well-designed workspace, there is also a culture and habit that indirectly affects the use of interior space functions and the atmosphere in the workspace. Based on the background above, this research is important for identifying issues and implementing well-designed space with cultures, habits, personalization and psychological issues for private workspaces in Surabaya City. The case study for this project is portfolios by students Regina, Aileen and Leony from the Interior Design Department – Petra Christian University.

2   RESEARCH METHODOLOGY

In general, the research method used is a qualitative research method. The primary data needed is in the form of data of the case study: Project 1 by Regina, Project 2 by Aileen, and Project 3 by Leony. The primary data will include the field data (existing interior conditions, user profiles) and a literature study regarding well-designed space. The methods used to collect primary data are direct observation, interviews, and literature studies.

Analysis was carried out by comparing literature data with field data. The thinking patterns used are deductive and inductive thinking patterns. In this study, comparative analysis using deductive thinking is analyzed using theories about the concept of well-designed space, the concept of space and its architectural, interior and cultural elements (including habit and activity), a d symbolic meaning of the sign in context. The analysis that uses inductive thinking patterns is analyzed using researchers' interpretations based on cultural contexts. Then, both these mindsets are combined to obtain optimal results. From this analysis, a conclusion can be drawn about the implementation of well-designed space in the directors' room in Surabaya (indicated as project 1, project 2 and project 3). The solution will be shown in the design for these projects.

3   LITERATURE

The theory used in this research is based on Augustin, Frankel and Coleman (2009) and the five aspects of a well-designed workspace: communicating, comforting, complying, challenging, and continuing.

3.1   *Well-designed space – communicating*

Communicating here means to communicate to others who we are, and at the same time to remind ourselves of who we are. In this particular aspect, workspace design should be used by users to communicate who they are as individuals, and at the same time remind themselves of this The user can even personalize (add useful or decorative objects to) their workspaces to accomplish the same things, within the bounds established by their employers and the culture it rewards. At the same time, employees can also try to understand what their colleagues are saying through their workspaces and what their company is trying to communicate about its values through the physical form of its workspaces. The physical form of the workplace can communicate a general level of concern for worker well-being.

3.2   *Well-designed space – comforting*

Comforting design is when the workspace can make people choose positive or negative moods. Color can communicate and influence the mood of workers. For example, red will make a person's mood negative. Saturation and brightness of the color can also be predicted and has the potential to produce negative effects. People will be stressed when their work environment doesn't say what they want. Stress also puts people in a bad mood and reduces motivation

3.3   *Well-designed space – complying*

Complying means meeting the need of increasing productivity and work motivation. Productivity is by increased via a few aspects regarding physical needs. For example, having privacy while working, a natural environment (natural lighting, view to outside, plants, aquariums, arts), a room temperature around 24 degrees, no unwanted noise, walls in an energetic yet calming color.

3.4 *Well-designed space – challenging*

Challenging means having an opportunity to grow and develop as a person. A workspace can make people in a better mood, which broadens their thinking, but if individuals in a space do not have a particular specialization or motivation, they will not be creative. Therefore, there will be creative spaces and the wall color should be warm colors and tones.

3.5 *Well-designed space – continuing*

Continuing means recognizing that a workspace will be passed from one generation to the next., for example, workspace design that can be used for the baby boomer generation through to the gen Z.

4 RESULT AND DISCUSSION

For this research, three individual directors' room became the focus of analysis. The design solutions were proposed by Regina, Aileen and Leony, students from the Interior Design Department – Petra Christian University.

4.1 *The user profiles and workspace issues – problem*

The user profiles used in this research are shown in Table 1. There were three projects with three explanations: existing workspace conditions, user profiles and workspace problems.

Table 1. User profiles.

Project name	Existing workspace condition	User profile (activity, character, habit and culture)	Workspace problems
Project 1 : By Regina	Existing condition 	User profile 	Layout location
	• Located in the middle between the meeting and accounting rooms. • Has two windows (to production and circulation area)	• User is from Generation X • Activity: workspace task and marketing. • Favorite item: plants and horse	• Workspace is lacking a sense of privacy (most walls are made of glass) • There is little drawer storage • No outlet on work desk

(Continued)

51

Table 1. (*Continued*)

Project name	Existing workspace condition	User profile (activity, character, habit and culture)	Workspace problems
Project 2 : By Aileen	• The ventilation uses three air conditioners. • The Director and Assistant Director are in the same room. • The lighting is 6 TL Lighting. • There is small window, which connects with the accounting space	• Favorite colors: blue and silver • Interior style: modern, minimal, simple	• No plants • The position of director and assistant director is too close • The atmosphere color is monotone and cool atmosphere. • No natural light • No art
	Existing condition 	User profile 	Layout location
Project 3 : By Leony	• The design in the CEO room is less effective because the large space cannot be maximized by the arrangement of existing furniture and the room feels less private • Has two windows (to production and circulation area) • The ventilation uses air conditioners.	• User is from Generation Y • Activity: workspace task, meeting, work, eat and marketing	• Workspace is lacking a sense of privacy • Need space larger area to store • Need atmosphere that produces an uplifting working spirit. • There is little drawer storage • The colors and patterns used are too monotonous
	Existing condition 	User profile 	Layout location 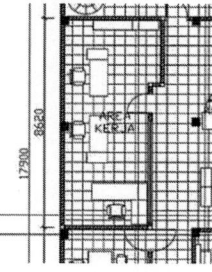
	• Located in the corner of the space • The ventilation uses air conditioners • The lighting uses downlights	• User is from Generation X • Activity: meeting, consult, discuss workspace task and marketing.	• Need a partition for the consultation room • The design needs to support the activity • Needs ergonomic furniture

(*Continued*)

Table 1. (*Continued*)

Project name	Existing workspace condition	User profile (activity, character, habit and culture)	Workspace problems
	• The color of the room is mainly cool color tones		• Needs decorations like plants and art ork • Lighting needs to support the work ambience • Need more private areas • Design is not suitable for next generation

Based on the workspace problems stated in Table 1, the concepts of well-designed spaced by considering users' culture and profile as design solution should be applied. The conclusion for the workspace problems will be privacy in the areas; a need for more space for storage and activities; a design that can last for the next generation; apply preferred colors and styles in the designs; more warm tone colors; natural lighting and ventilation; art, plants and decor for personalization.

4.2 *Solution*

There will be concepts and solutions for each project. The well-designed space concepts will be explained as in Table 2.

Table 2. Well-designed space concept by considering users' cultures and profiles.

Well-designed space concept	Project 1 by Regina (Figure 1)	Project 2 by Aileen (Figure 2)	Project 3 by Leony (Figure 3)
Communicating	• Simple design ➔ shows neat properties and design for Generation X. • High and elegant chair ➔ shows high rank • Separate space • Backdrop that is simple and high ➔ denotes special position and rank	• The desk design need to be good for communication between workers	• Design will increase concentration of the user • Design will increase communication within associates
Comforting	• Color based on user preference: for example using blue and grey as accents • Design based on user's wishes ➔ can work effectively, efficiently, and productively • Smells, light and noise at a comfort level	• The CEO's office will be more private • Neutral colors used that seem boring and less comforting	• Need private pantry with a café ambience • Use a warm color to create a more relaxing ambience • More space between two tables to create privacy
Complying	• Seating position at the back • Use natural lighting from skylight and light pipe	• The atmosphere in the workspace should be more open, comfortable and pleasant • More storage for important documents	• Design need to adjust for the next generation • The table will be made larger to adapt more technology

(*Continued*)

Table 2. *(Continued)*

Well-designed space concept	Project 1 by Regina (Figure 1)	Project 2 by Aileen (Figure 2)	Project 3 by Leony (Figure 3)
	• Create privacy using magic glass. • Art installation of horse figure and plants on table and skylight • Using the user's favorite color on the walls (blue, grey, white) • View to outside and using a big window.		• More storage for filling near the user
Challenging	• There is creative space • Using plants in the room • Using warm tones for creative thinking	• Increasing the privacy of each room, so that each occupant in the workspace is not disturbed by others workers	• Using warm tone and natural colors and textures to stimulate creative thinking • Using plants to create more relaxing areas
Continuing	• Creating an adaptable design for the next generation • Using technology for the design that can be used for 5–10 years	• Design needs to meet standards • Need to add space to increase the number of workers • Add a pantry, and also a baby changing place in the bathroom	• Design should be able to adapt fo the next generation and be used for 5–10 years.

Each of the concepts will be applied in the project interior designs as the design solutions for a well-designed space concept. The design solutions can be seen in Figures 1–3 below.

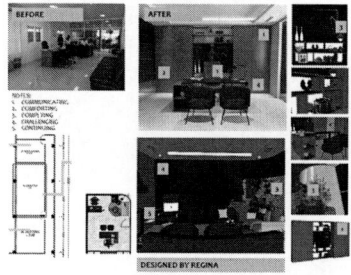

Figure 1.   Well-designed space concept of Project 1 by Regina.

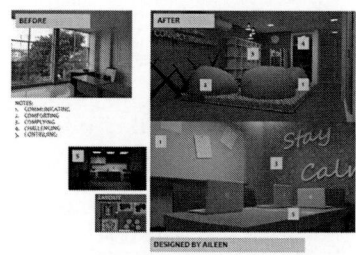

Figure 2.   Well-designed space concept of Project 2 by Aileen.

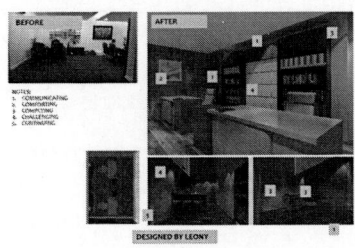

Figure 3. Well-designed space concept of Project 3 by Leony.

Based on the well-designed space concept and design solutions in Table 2, the general concept is more private areas using high chairs and partitions; more space for storage by adding furniture; more space for meetings and group discussions; the design can be used for the next generation by adding additional technology to the furniture; implementation of favorite colors and styles (for example blue color in Project 1); more warm toned colors by adjusting light color, adding chocolate colors and wood textured; adding natural lighting and ventilation using windows, light pipes or skylights; and adding art installation, plants and decoration based on users' favorite items.

5   CONCLUSION

From the results of the three case studies, it can be concluded that a good workspace has a good influence on its users. Well-designed space will influence users by make them happy and ultimately improve work performance and efficiency. The implementation design of well-designed space and psychological concepts were needed in order to increase users' working performance.

From these three case studies, where the workspace was previously very plain, monotonous and less comfortable, it has become a more comfortable and fresh design in accordance with the character and user profile. This design will help workers improve their work performance. For example, previously there were no plants in the workspace, but these could provide a sense of comfort, and creativity. Another example is personalization for the users using art installations in the workspace. In conclusion, as an interior designer we must be able to understand who the user of the space is and create a space that is physically and psychologically good and in accordance with the characteristics and needs of the user.

ACKNOWLEDGEMENTS

This study was supported by the Interior Design Department, Faculty Art and Design, Petra Christian University Surabaya-Indonesia, and by family. The authors are grateful for this support.

REFERENCES

Augustin, S., Frankel, N. and Coleman, C. (2009) *Place advantage: Applied psychology for interior architecture*. John Wiley & Sons.
Ishananto, O. (2010) *Mal: surga tanpa Tuhan, ruang tanpa waktu*. Jaring Pena.
Van Meel, J., Martens, Y. and van Ree, H.J. (2010) *Planning office spaces: a practical guide for managers and designers*. Laurence King London.
Sari, S.M. (2018) 'Makna Ruang Interior Shopping Mall 'Tunjungan Plaza'Di Surabaya'. PPS ISI Yogyakarta.
Snyder, C.J. and Anthony, J. (1991) *Pengkajian Lingkungan Perilaku by Gary Moore, Introduction to Architecture*. Jakarta: Penerbit Erlangga.

Emerging Trends in Psychology, Law, Communication Studies, Culture, Religion,
and Literature in the Global Digital Revolution – Setiawan & Rahmawati (eds)
© 2020 Taylor & Francis Group, London, ISBN 978-1-03-224216-3

Small industry and the relationship with protecting and managing the environment

Dewi Tuti Muryati, Endah Pujiastuti & Tri Mulyani
Universitas Semarang, Semarang, Indonesia

ABSTRACT: Economic development is an important instrument in driving the dynamics of the
Indonesian national economy. In its industrial development policy, the Indonesian nation has
a commitment to developing its industrial sector as an important part of the constellation of
national development. Industrial development is directed towards the independence of the national
economy, increasing competitiveness, and increasing market share at home and abroad by always
protecting the environment. In order to carry out sustainable national development, industry players should not neglect government policy in the form of a set of legal regulations aimed at preserving environmental functions, such as Law Number 32 of 2009 concerning Environmental
Protection and Management, Law Number 3 of 2014 concerning Industry, Government Regulation Number 27 of 2012 concerning Environmental Permits, Government Regulation Number
107 of 2015 concerning Industrial Business Licenses, and other legal regulations. There is an obligation towards the adherence of small industry actors to environmental propositions and the rule
of law as an instrument for preventing pollution that is a result of industrial activities.

Keywords: small industry, regulation, environment

1   INTRODUCTION

A development characteristic of developing countries is placing the industrial sector as an
alternative to advance the economy. In other words, the role of the industrial sector is considered to greatly support the success of the country's economic development, especially in the
long-term development perspective. Based on this assumption, the World Commission for
Environment and Development reports that industry occupies a central position in the economy of modern society and is an irreplaceable driving force for growth (Faishal, 2016). Indonesia as a developing country, is still designing its national development. One of the national
development priorities is development in the economic field by increasing the competitiveness
of national industries through efficiencies and building competitive advantage as a foundation
for strengthening economic resilience and growth. Several industrial forms are the foundation
of national development, The acceleration of the industrial sector is a real effort to grow and
develop industrialization in the country (Brutland, 2018). The role of small and medium enterprises is very important because in the current and future developments the manufacturing
industry sector will be dominated by small and medium industries.

Although industrial development policy has an important role in national development,
applying it has a negative influence on the environment, namely the emergence of industrial
waste pollution which threatens the sustainability of environmental functions in the long
term. In order to carry out sustainable national development, small industry players should
not simply ignore government policies in the form of a set of legal regulations aimed at the
preservation of environmental functions, such as Law Number 32 of 2009 concerning Environmental Protection and Management, Law Number 3 of 2014 concerning Industry, and
Government Regulation Number 27 of 2012 concerning Environmental Permits. There is an

obligation for the adherence of small industry actors to environmental propositions, and the rule of law as an instrument for preventing pollution. Therefore, a more in-depth study needs to be carried out to analyze efforts to regulate small-scale industries in relation to environmental protection and management (Undang-Undang Nomor 32 Tahun 2009; Peraturan Pemerintah Nomor 27 Tahun 2012, 2012; Undang-Undang Nomor 3 Tahun 2014)

2 DISCUSSION

2.1 *Environmental management policies*

Before further elaborating on the substance of environmental management policies, it is first necessary to explain the definition of "policy". Etymologically, policy or wisdom is an Indonesian translation of Dutch legal terms, namely *politiek* (which includes the meaning of *beleid*) from the word *rechtspolitiek*, which is a form of two words *recht* and *politiek* (Termorshuizen, Supriyanto-Breur and Djohan-Lapian, 1999).

Drupsteen gives the understanding that a policy contains the overall objectives and means of certain actors. In simple terms, a policy is the answer to one question about what is achieved by someone, through how, and by what means it is implemented (Bethan, 2008). From this opinion, the understanding is that the policy contains elements of purpose, goals, and objectives to be achieved or aspired to.

Environmental management policy is state or government policy in the field of environmental management. The policy has specific goals and objectives on how and by what means environmental management is carried out to achieve these goals and objectives. Environmental management policies are also closely related to laws or regulations and lay the basic framework for public policy. Based on these arguments, environmental management policies are established and carried out based on the provisions of the law or legislation.

The general explanation of Law Number 32 of 2009 states that environmental management must be able to provide economic, social and cultural benefits that are carried out based on the principles of prudence, environmental democracy, decentralization, and recognition and respect for local wisdom and environmental wisdom. Furthermore, environmental protection and management requires the development of an integrated system in the form of a national policy on environmental protection and management that must be carried out in a consistent and consistent manner from the center to the regions. For this reason, the state is responsible for ensuring the goals of environmental protection and management can be achieved to realize the welfare of the population. Therefore, we need the support of legal instruments, in the form of laws and regulations in the field of the environment, that are capable of realizing Indonesians as environmental advisers.

2.2 *Overview of industry*

Industrial activities have an impact on various aspects of people's lives, as well as resulting in complex social problems such as industrial waste, which will ultimately disrupt aesthetics and effect a decrease in environmental quality. Industry is a business activity to change goods or process raw materials into finished goods that are ready for use by those who need them. In Article 1 number 1 of Act Number 3 of 2014, industry is an order and all activities related to industrial activities. Furthermore, in Article 1 point 2 of Act Number 3 of 2014, industry is all forms of economic activities that process raw materials and/or utilize industrial resources to produce goods that have higher added value or benefits, including industrial services.

Based on the definition of the Badan Pusat Statistik, small industries are business units in the manufacturing sector that do not use labor and that employ between one and 19 people. Furthermore, small industries can be classified into two sub-sectors according to their workforce: home industries (cottage industries or household industries), namely business units without workers or with between one and four employees; and factories/small factories or small workers, namely business units with a workforce of between one and 19 people.

The development of the industrial sector as part of the national development process in increasing economic growth has brought changes to people's lives. These changes include the impact of industrial development on the socio-economic community and the environment in the vicinity. The impact of industrial development on socio-economic aspects includes the livelihoods of the population, and the offering of wider employment opportunities. The impact of the industry on socio-cultural aspects includes the diminishing power of binding cultural values and norms. The impact of industrial development on the environment can have a negative influence on people's survival.

2.3 *Environmental impact of small industrial activities*

In addition to producing desired products, industrial activity also creates various types of waste and can cause damage to the environment. Pollutants that enter the environment will interact with other environmental components, environmental balance can be disrupted and the quality of the environment also changes. In accordance with one of the industrial development goals as stated in Article 3 letter c of Law Number 3 of 2014, the aim is to create an industry that is independent, competitive and advanced, as well as a green industry. The existence of industrial activities is expected to provide added value from the use of resources by not causing damage to their environment. Environmental management in the industry has undergone a paradigm shift from the Atur Dan Awasi (ADA) paradigm to the Atur Diri Sendiri (ADS). The ADS environmental management system places greater responsibility on the community to maintain compliance with regulations and there is widespread public pressure on the industry to be environmentally friendly. Threats to the existence of industry cause the industry to carry out environmental management which gives freedom to regulate themselves. The sustainability of industry in the era of free trade determines that it is not only oriented to economic benefits but also responsible for the environment and society.

Small industries consider that their waste output is very small while the operational costs in environmental management are felt to be quite burdensome because they will increase production costs. Environmental pollution from small industries is generally caused due to the use of simple equipment and technology, inefficient production, poor governance and financial inability to control pollution. Eco-efficiency is an instrument of the ADS environmental management system by reducing production costs from processes that are not needed so as to minimize environmental pollution.

2.4 *Regulation of small industries in relation to environmental protection and management*

Changes to Law Number 5 of 1984 into Law Number 3 of 2014 carry the mission of changing the new paradigm of "Green Industry". Based on Article 1 point 3 of Act Number 3 of 2014, green industry is an industry which in its production process prioritizes efforts to efficiency and effectiveness of sustainable use of resources so as to harmonize industrial development with the preservation of environmental functions and benefit to the community.

The government's attention to development in the industrial sector is reflected in Article 9 paragraph (1) letter b of Law Number 3 of 2014, which states that the National Industrial Development Master Plan is prepared with the least attention to an industrial culture and local wisdom that grows in the community. From these provisions, it can be seen that the mainstream of this century's industry must return to the things that are in the nature of preserving the environment because it places the industry as a place of production to produce capital, not for income.

The linkage of the green industry paradigm (Law Number 3 of 2014) to environmental regulations (Law Number 32 of 2009) can be seen from the provisions concerning industrial empowerment contained in CHAPTER VIII of the First Section on Industry Small and Medium Industries, especially Article 72 paragraph (2) and Article 75 paragraph (1).

From these provisions, then it will be a derivative norm for the establishment of policy regulations in which the government plays an active role in directing and controlling industrial activities through environmental administration facilities. Furthermore, it can be stated that for small and medium industries there are special arrangements relating to environmental policies as formulated in Article 32 of Law Number 32 Year 2009.

The Industrial Law emphasizes the existence of a commitment to developing a green industry as a work culture for the industrial workforce. Implementation will be through a clean production process and reduce, reuse, reprocess, and restore, or known as the 4Rs (reduce, reuse, recycle, recovery).

In addition to the green industry paradigm, in order to achieve industrial growth, the licensing aspect also plays an important role. Licensing is one of the tools of wisdom which, if used efficiently, will be an effective tool to drive the development of the business world into fields that truly support development. Based on Article 3 paragraph (1) of Government Regulation Number 107 of 2015 concerning Industrial Business Licenses that industrial business licenses (IUI) include IUI for small industries, medium industries, and large industries.

An industrial business permit, called IUI, is also essential to protect the environment from pollution and its position alongside the Trading Business License (SIUP). Certain industrial companies in small industry groups that are excluded from having IUIs are given a Small Industry Registration Certificate (STPIK). The relationship between IUIs and STPIKs with the environment is inseparable and is the essence of licensing. Thus, the provision of IUI and STPIK is one of the administrative juridical means of preventing environmental pollution. The regulation, guidance, and supervision of industrial activities are one of the efforts to realize environmentally sound development.

3 CONCLUSION

In accordance with one of the goals of industrial development, namely to improve the prosperity and welfare of the people fairly and evenly by utilizing funds, natural resources, and/or aquaculture products and by paying attention to the balance and preservation of the environment, the existence of industrial activities is expected to provide value-added use of resources by not causing damage to the environment. Likewise, with the existence of small industries that contribute to environmental damage in the form of pollution, it is necessary to make arrangements through government policies in the form of the following regulations:

a. UU no. 32 of 2009 concerning Protection and Management of the Environment, namely in the form of analisis mengenai dampak lingkungan (amdal) preparation assistance for businesses and/or activities of weak economic groups that have an important impact on the environment and facilitation, costs, and/or analisis mengenai dampak lingkungan (amdal) preparation.
b. Law Number 3 of 2014, namely the green industry paradigm in the framework of empowering industries, including small and medium industries.
c. PP No. 107 of 2015, namely in the form of a Small Industry Registration Certificate (STPIK).
d. Government Regulation Number 27 of 2012, which is in the form of a Declaration of Environmental Management and Monitoring (SPKPPL).

REFERENCES

Bethan, S. (2008) *Penerapan prinsip hukum pelestarian fungsi lingkungan hidup dalam aktivitas industri nasional: sebuah upaya penyelamatan lingkungan hidup dan kehidupan antar generasi*. Alumni.
Brutland, G. H. (2018) *Hari Depan Kita Bersama*. Jakarta: Gramedia.
Faishal, A. (2016) *Hukum lingkungan: pengaturan limbah dan paradigma industri hijau*. Pustaka Yustisia.
Peraturan Pemerintah Nomor 27 Tahun 2012 (2012) *Izin Lingkungan*.
Termorshuizen, M., Supriyanto-Breur, C. and Djohan-Lapian, H. (1999) *Kamus Hukum Belanda-Indonesia*. Djambatan.
Undang-Undang Nomor 3 Tahun 2014 (2014) *Perindustrian*.
Undang-Undang Nomor 32 Tahun 2009 (2009) *Perlindungan dan Pengelolaan Lingkungan Hidup*.

Emerging Trends in Psychology, Law, Communication Studies, Culture, Religion, and Literature in the Global Digital Revolution – Setiawan & Rahmawati (eds)
© 2020 Taylor & Francis Group, London, ISBN 978-1-03-224216-3

Redefining the relationship of majority and minority as a social principle

J. Basyir
Alauddin State Islamic University of Makassar, Makassar, Indonesia

R.F. Marta
Universitas Bunda Mulia, North Jakarta, Indonesia

Y.B. Setiawan
Universitas Semarang, Semarang, Indonesia

ABSTRACT: This study aimed to reconstruct the definition of the majority and minority relationship from Jurgen Habermas' perspective, which is always seen from the aspect of dominat and being dominated. What has strongly shown up in this study is that communication becomes a praxis that can make every single subject involved in the agenda for social change. It elucidates that both majority or minority people must realize that they are part of human society and whatever the interest, perspective, or status, they must live in a social community. This can be accomplished by constructing many experiences of communication without any domination, monopoly, arrogance, intimidation, subjectivity, and exclusivity in a public sphere.

Keywords: majority, minority, public sphere, social community

1   INTRODUCTION

Redefining the relationship between majority and minority groups must be seen as a matter of urgency, although it has the potential to be a challenge for a majority group who may feel desolated by this kind of notion. Nevertheless, this concept must be discussed considering that the majority group always constructs an unfair and hierarchical intercourse. This thought is strongly confirmed by the perspective of anti-essentialism, which indicates that certain elite classes have used their power to command and promote their interests through cultural instruments (Barker, 2005).

This problem is no less injurious to humans as social beings (Raditya, 2017) who want their existence to always be socialized. Explicit in this proposition is that there is still a community group that with a surplus of assets that tends to show its superiority toward other groups of people who are far behind the majority group even though those attitudes and actions do not reflect the meaning of social values. Social antagonism (Jørgensen and Phillips, 2007) is another concept that describes how a different group society seeks to abolish another group in society. However, polarization must not always be understood as a binary or antagonistic opposition relationship in which there is an attempt to legalize and legitimize a group's position structurally.

At this level, humans cannot destroy their own character to enjoy and affiliation with other humans or to group themselves into other societies because after all humans are nonmonolithic or multifaceted beings (Samovar et al., 2017), which is the tendency of humans to form community groups affiliating with certain interests, goals, and ideologies that influence them in the middle of a majority and dominant group of people.

With this perspective, it is important to underline that nonmonolithic issues in human relations should not be used as the arena of the practice of power where polarization exists and to keep them from being treated or only arbitrarily articulated. George Wilhelm Friedrick Hegel (Takwin, 2003) mediates the tendency of this relationship with his dialectical idea. Simply stated, the purpose of dialectics is that a human entity or a community group must require the presence of human life or other society in order to create mutual relations.

In the context of the philosophy of communication, Martin Buber called it as reciprocal relations. What Martin Buber wants to ascertain is emptiness or the absence of humans driven by other humans who only view these human beings as an I-it relationship, so that a person is treated according to the type or category attached to him, for instance, factory workers or ethnic minority. There is nothing wrong with evaluating other humans, but what is missing is when human actuality is ignored by social systems, interests, ideologies, et cetera, which in turn excludes human existence. In a simple view, Buber only wants to create healthy, harmonious, and humanizing relationships through interaction (Buber, 1961).

Inclusive awareness and the principle of impartiality are the manifestation of what we call social beings, not a partiality awareness that tries to look at and treat humans from the characters and categories they have. Social beings are creatures who have a dimension of dependence on other creatures, so that they have to not harm other people's feelings. As an illustration, when an entrepreneur wants to operate and develop his business, he must work together with his employees. Conversely, when employees need money for daily life, they must also work on the situation of their boss's business activities. It is this phenomenon that explains how the dialectical relationship is displayed as a dialectical awareness or reciprocal relation (Juliardi, 2014).

The preceding illustration finally leads us to a discussion about the fact that humans are not monolithic and it then simply establishes an ambition of power and domination. According to the Economist Intelligence Unit (EIU), Indonesia is ranked from 68th (2017) going down to 48th (2016) in democratic system index implementation (Peterson, 2018). Similarly, the Wahid Foundation Institution director Yenny Wahid illustrated that there were around 200 incidents of violations of religious freedom aimed at minority groups including fellow Muslims as the majority religion in Indonesia until 2017, which was dominated by different treatment when accessing public facilities and barriers to performing worship rituals (Harvan, 2017). The data emphasize exactly how the democratic system is not running well, mostly in the behavior of majority groups toward minority groups, and lucidly makes clear that democracy, including the state (government), does offer domination and hegemony for those majority groups.

The diversity of ideology, customs, or religious views that crystallize majority and minority relations should not be a reason for humans to not respect one another. It is not ethical enough to understand the duality of public relations, especially considering this as a threat to humanity as social beings. It means that humans can still be social beings even though they have different views, as being different is a sure thing. The strength and power of a society are not determined by individuals themselves, but by communities (Uhi, 2016).

2 RESEARCH METHOD

This study used a critical paradigm design that focused on Jurgen Habermas' perspective. His method concentrates on rational communication that illustrates how the act of communication is able to provide more occasions for emancipation and is more progressive toward individual autonomy in the established social system (Hardiman, 2009).

3 RESULTS AND DISCUSSION

The research of Agustini (2015), Student of Communication Sciences, Alauddin State Islamic University of Makassar entitled *"Relationship Acceptance between Javanese Ethnic and*

Mandar Ethnic in Wonomulyo Sub District, Poliwali Mandar District, West Sulawesi (Phenom-enology Study)" (Agustini, 2015), proves that there is social and cultural understanding between Javanese and Mandar ethnic groups. The acceptance between the two ethnic groups could be achieved because of the similarity of Islamic religious ideology, cultural interest, work ethic of Javanese ethnicity, and politics.

Based on Jurgen Habermas' perspective, these are not just the natural activities found in Ika Agustini's research, but are mostly about the idea of rational communication particularly proposed by the minority Javanese ethnic group. The public sphere as a social facility is proven to be able to dilute the atmosphere of freezing relations between the majority and minority groups represented by each Javanese and Mandar ethnic group. Radical democracy, the modern term used by Jurgen Habermas, is a form of social relations that takes place within the scope of free mastery of communication in the sense that the minority class struggle in the form of conflict or political revolution is changed with rational conversation as the main element of change or emancipation. In a simple sense, the change can be achieved through communicative actions that are oriented toward achieving mutual understanding as a form of success in building social relations (Hardiman, 2009).

Another foundation that may be built from the events of the Javanese and ethnic Mandar meetings is human relations based on social exchange relations. George Casper Homans in Moch Syahri (2014), for example, explains that human relations are built on the basis of getting a lot of rewards and minimizing penalties (Syahri, 2014). This is clearly indicated when ethnic Javanese entered the area which is now known as Wonomulyo subdistrict, bringing some skilled workers headed by R. Soeparman, among whom were teaching staff, health workers, agricultural workers, and land workers (Sunani, 2013). The reason for the presence of these experts is the efforts of Javanese people to be accepted by the indigenous Mandar community because these events not only expose exchange events or social transactions, but also indicate the effort of building up social equality through those social transaction, so that the communication will be more humanistic.

The concept of Jurgen Habermas critically argues the typology of the majority as a society that tends to consider themselves as normal people (the protagonists), full of power and of higher degree. Indeed, what happened to them is not merely an event of social transactions whose results are in the form of ethnic Mandar acceptance of ethnic Javanese, but there is a kind of awareness of political actions to attain greater existence and to have the same social occasion through the public sphere.

This study firmly would like to illustrate the gap that can be seen from Jurgen Habermas's perspective, focuses greatly on rational communication. What the researcher clearly found is that it is not only about the same attributes from those groups or discontinuity of the majority's typologies, but it is mainly more about rational communication in the public sphere. It definitely indicates that what is officially built up in that public sphere is existence and acknowledgment either as the community or human beings through communication experiences between them. This activity is in fact constructing the awareness of their inclusivity. Jurgen Habermas believes that it is not a revolution or conflict that will reduce unequal social relationships, but rather the activity of communication. However, it all can be done only if the public sphere is free from any tyranny government or certain fanaticism, so that it can go on with humanity and a very good situation.

Based on the preceding discussion, the author wants to assert the position of Jurgen Habermas who gives more attention to the power of the communication paradigm through its formal construction by using a consensus path that creates the development of *"Radical Democracy,"* a social order that promotes communication without any authorization. Domination, monopoly, arrogance, intimidation, subjectivity, and group or self-exclusivity should be removed to create an ideal conversation situation. In summary, public sphere policy that requires the recognition of and respect for each other's groups provides an atmosphere conducive to creating a new pattern or culture that is more civilized and humane in solving every humanitarian problem.

4 CONCLUSION

Commitment and consistency are the two main social modes for developing an equal relationship between majority and minority groups as demonstrated by the Javanese minority and Mandar majority since 1973. If the commitment and consistency are still declining in the society, it is not possible then to produce the essence of communication which is equality. Equality does not mean an ability to produce an identical message, but the most essential is to consider other participants as equals, which acknowledges existence. In other language, the equality produces the existence. The more people equalize other people, the more existence they have. It explains that communication is not about looking for whether something is true or false, but it is mostly to achieve the same perception through consensus. In the other perspective, communication experiences promote any groups to develop and assist each other as social communities. Moreover, current communication is not limited to distance, time, and space. Communication can occur anywhere, anytime, without requiring face-to-face contact. Even the presence of social media is able to negate social status (majority or minority), which is often seen as an obstacle to communications (Watie, 2016).

REFERENCES

Agustini, I. 2015. *Relasi Penerimaan antara Etnis Jawa dan Etnis Mandar di Kec. Wonomulyo, Kab. Polewali Mandar, Prov. Sulawesi Barat (Studi Fenomenologi)*. Universitas Islam Negeri Alauddin Makassar.

Barker, C. 2005. *Cultural Studies: Teori dan Praktik*. Yogyakarta.

Buber, M. 1961. *Between Man and Man*, trans Ronald Gregor Smith. London: Kegan Paul, p. 176.

Hardiman, F. B. 2009. *Kritik Ideologi, menyingkap Pertautan Pengetahuan dan Kepentingan bersama Jurgen Habermas*. Yogyakarta: Penerbit Kanisius.

Harvan, M. 2017. *Yenny Wahid: Buang Jauh-Jauh Sikap Tirani Mayoritas atas Minoritas*, rappler.com. Available at: http://www.google.com/amp/s/amp.rappler.com/indonesia/berita/177364-yenny-wahid-tirani-mayoritas-minoritas (accessed February 18, 2019).

Jørgensen, M. & Phillips, L. J. 2007. *Analisis Wacana: Teori & Metode*. Pustaka Pelajar.

Juliardi, B. 2014. *Ilmu Sosial Budaya Dasar*. Bandung: Alfabeta.

Peterson, D. 2018. *Indonesia's Minority Report*. New Mandala, p. 14.

Raditya, I. N. 2017. *H.O.S. Tjokroaminoto Memadukan Islam dan Sosialisme*, tirto.id. Available at: http://tirto.id/hos-tjokroaminoto-memadukan-islam-dan-sosialisme-cwW1.

Samovar, L. A. et al. 2017. *Communication between Cultures*. Nelson Education.

Sunani, U. 2013. *Kampung Jawa di Poliwali Mandar*.

Syahri, M. 2014. *Teori Pertukaran Sosial George C. Homans dan Peter M. Blau*. Surabaya: Universitas Airlangga Press.

Takwin, B. 2003. *Akar-akar Ideologi: pengantar Kajian konsep Ideologi dari Plato hingga Bourdieu*. Penerbit Jalasutra.

Uhi, J. A. 2016. *Filsafat Kebudayaan*. Yogyakarta: Pustaka Pelajar.

Watie, E. D. S. 2016. Komunikasi dan Media Sosial, *Jurnal The Messenger*, 3(2), p. 69–74.

The scooter motorbike market monopoly by Honda and Yamaha in Indonesia, 2016–2019

Rr Chairunnisa Windiatama Putri
Universitas Indonesia, Indonesia

ABSTRACT: Analleged cartel case with a price-fixing scheme is being handled by the Business Competition Supervisory Commission (KPPU) in Indonesia. PT Yamaha Indonesia Motor Manufacturing (YIMM) and PT Astra Honda Motor (AHM) are suspected of regulating the sales price of 110–125 cc automatic scooters in 2013–2014. The court case was contested fierce because AHM and YIMM rejected the allegations, but after strong evidence goiven in the hearing and the Supreme Court Verdict Number 04 / KPPU-I / 2016, PT AHM and PT YIMM must comply with the law. The law that has been violated is Article 5 Paragraph 1 of Act Number 5 of 1999 concerning Prohibition of Monopolistic Practices and Unfair Business Competition. AHM and YIMM are suspected of carrying out cartels that can harm consumers. A request for cessation submitted by AHM and YIMM was rejected by the Supreme Court. Thus, the case that KPPU started in 2016 has been continuously in court until now, in 2019.

1 INTRODUCTION

In the case of scooter pricing, the Business Competition Supervisory Commission (KPPU) capitalized on several e-mails from one of the motorbike manufacturers that was allegedly part of a price conflict between the two Japanese motorcycle brands. Based on KPPU investigators in the last follow-up hearing on January 5, 2017, suspecting that there was a meeting between the management of PT YIMM and PT AHM discussing the price agreement. According to KPPU, Yamaha would follow Honda's selling price, which was then followed up with an e-mail leading to an increase in Yamaha scooter prices to follow Honda's prices. The problem is that this e-mail evidence is debated by Yamaha and Honda, who from the outset denied the cartel. The Chairperson of the Indonesian Motorcycle Industry Association (AISI), Mr. Gunadi Sindhuwinata, stressed that the email, which became the basis of the KPPU case, ensnared Honda and Yamaha: it was only an internal discussion and had nothing to do with price agreements. In addition, the issue of motorbike price movements that go on between Honda and Yamaha is only a strategy of watching competitors when both have to raise prices due to factors such as wage increases and others. Mr. Gunadi said that this is only a matter of courage about who first prices raised (tirto.id).

Yamaha and Honda are both engaged in the transportation sector and the product discussed here is scooter motorcycles. In 2016, Honda and Yamaha were very close in terms of sales, with Honda slightly outperforming Yamaha. However, by September 2016, Honda's sales reached 423,000 units while Yamaha only sold 119,000 units. With such a wide difference in sales figures, Yamaha then moved their prices closer to those set by Honda. However, if Yamaha followed Honda in price, then price fixing occurs because it can then be assumed that other companies will follow Yamaha and Honda and that price is considered the actual price.

Monopoly is a market control carried out by a person, company or entity to master market offerings (sales of goods and or services on the market) aimed at its customers. When there are two companies that work together in monopolizing the market (duopoly), cooperation is carried out in terms of pricing. Based on the decision of the case issued by the Supreme Court number 04 / KPPU-I / 2016, there are sub-chapters that discuss price-fixing, it said price fixing needs not be the same nominal number; the essence of Article 5 of Law No. 5 of 1999 is that there is no restriction to set the price but it is prohibited to make a deal to set the price. The chronology of pricing, as written in the Supreme Court's verdict Number 04/KPPU-I/2016 in the written sub-chapter, is that the first meeting of the presidents of Honda (Mr. Inuma) and Yamaha, (Mr. Kojima) was when they played golf in 2013. In January 2014, there was a further golf match between Honda and Yamaha directors. In April 2014, there is evidence of an internal Yahoo email from the president to VP of Marketing and VP of Forwarding to the Marketing Management Group. In November 2015, there was discussion of Honda's president (Mr. Inuma) playing golf with the president director of Yamaha (Mr. Kojima). In January 2015, there was an e-mail from witness Mr. Terada regarding the pricing issue from Mr. Kojima to Mr. Inuma. Based on the facts of the trial, the email sent on January 10, 2015 was a letter sent by Witness Mr. Yutaka Terada who at that time was the Marketing Director of Reported Party I using the email address teradayu @yamaha motor.co.id and sent to Dyonisius Beti as Vice President Director of the Reported Party I.

When competition is carried out in a healthy manner, the price should be fixed close to the price of production. When prices move down close to production costs, the market will be more efficient, and the effect will be to increase savings for consumers which can lead to consumers welfare (welfare improvement). But if a group of companies (oligopoly) fix a price, then it will rise well above the price of production. This will be detrimental to consumers. Evidence in the form of e-mails between the two companies found by KPPU contained negotiated price fixing and culminated in the meeting of the two companies on January 5, 2017. After the meeting there was an increase in Yamaha prices following the Honda price (reported in tirto.id).

2 LAW IN INDONESIA

As reported on the cnnindonesia.com website on February 21, 2017, the verdict of the assembly stated that YIMM and AHM was proven to have legally and convincingly violated Article 5 paragraph (1) of Law Number 5 Year 1999 concerning Prohibition of Monopolistic Practices and Unfair competition. Thus, YIMM and AHM are subject to administrative sanctions. The two reported parties were also required to pay a fine for committing a violation. YIMM was fined IDR 25 billion and AHM was fined IDR 22.5 billion. Mr. Syarkawi said, "As we discussed in KPPU, the verdicts were in the form of fines and administrative sanctions. For the fine, the convicted party must pay the penalty and deposit it directly into the state treasury in accordance with the amount decided by the assembly." He said that the two reported parties had to pay fines and a copy of the proof of payment of the fine must be reported and submitted to the KPPU.

Yamaha-Honda has been proven to price fix. Price fixing is a criminal act because it will harm consumers. According to Swedberg (2009), "As an example of organizational crime – that is, criminal behavior that benefits companies but not necessarily individuals – pricing is common in all industrialized countries and involves a large amount of money." A recent study of price regulation showed that the social structure of this type of activity is suitable for network analysis (Baker and Faulkner 1993). Pricing of standard products (for example, switchgear) usually leads to a decentralized network, because a little direction is needed from above, while the reverse applies to more complex products. The more links to an actor in the pricing network, the greater the risk.

Article 5 Year 1999 prohibits business actors from conducting price fixing in Indonesia. It is stipulated in Article 5 of Law Number 5 Year 1999, as follows:

1. Business actors shall be prohibited from entering into agreements with their business competitors to fix the price of certain goods and or services payable by consumers or customers on the same relevant market.
2. The provisions intended in paragraph (1) shall not be applicable to the following: a. an agreement entered into in the context of a joint venture; or b. an agreement entered into based on the prevailing laws.

As written there, AHM and YIMM have violated the relevant regulations regarding price fixing, with e-mail as evidence. It was reported on cnnindonesia.com on December 8, 2017, that both Yamaha and Honda appealed the KPPU's decision to the North Jakarta District Court. But the judges still refused to cancel the decision. This caused the two firms to take cessation steps to the Supreme Court (MA). The Indonesian Consumers Foundation (YLKI) encouraged Yamaha Indonesia Motor Manufacturing (YIMM) and Astra Honda Motor (AHM) to first correct the automatic scooter sales price. According to the Deputy Daily Chairperson of YLKI Mr. Sudaryatmo, there are currently two court decisions stating that the two agreed to conspire about scooter sales prices in Indonesia.

In 2019, The Supreme Court rejected the appeal request of PT Yamaha Indonesia Motor Manufacturing (YIMM) and PT Astra Honda Motor (AHM). The Supreme Court's decision regarding this case was stated in Registration Number 217 K/PdtSus-KPPU/2019 on April 23, 2019. The total fine was IDR 47.5 billion: Yamaha was fined IDR 25 billion and Honda was fined IDR 22.5 billion (cnnindonesia.com).

3 CONCLUSION

The essence of Article 5 of Law No. 5 of 1999 there was no prohibition on setting prices, but what was forbidden was making agreements to set prices. Two decisions from different levels of trial stated that Yamaha and Honda had reaped excessive profits, which led to consumer losses. The cartel made excessive profits but caused consumers to lose money because they have to pay more for the product.

The author's suggestion to the KPPU and the Supreme Court is to impose strict sanctions (fast and right on target) on AHM and YIMM, as well as ordering a drop in prices by AHM and YIMM for their products. In essence, price cartels or price fixing will harm consumers. To pay for consumer losses, AHM and YIMM should reduce prices and pay fines to the state as determined.

Until now, the author still has not found a copy of the Supreme Court verdict number 217, so the new information is based on the news. On May 9, 2019, the Commission conducted a journalist forum on this case, with Commissioner Mr. Guntur Saragih (Chair of the National Consumer Protection Agency) (BPKN), Rizal E Halim (Advocacy Commission), and Gopprera Panggabean (Director of KPPU Enforcement).

Guntur Saragih explained that his party was awaiting a copy of the decision, Case Number 217 K/Pdt.Sus-KPPU/2019. In the end, the Supreme Court had rejected the appeal from YIMM and AHM against the KPPU's decision. Based on case information on the MA website, the 'reject' decision was declared on April 23, 2019. Guntur explained that KPPU would conduct a trial in the nearest court based on the domicile of the company concerned after receiving a copy of the decision from the Supreme Court. In the meantime, this case has not been completed and is still in the trial.

REFERENCES

Swedberg, R. (2009) *Principles of economic sociology*. Princeton University Press.
Gumilang, Prima. 2017. KPPU: Yamaha-Honda conspires to play automatic scooter prices. Accessed on December 26, 2018, from https://www.cnnindonesia.com/nasional/20170221002743-12-194853/kppu-yamaha-honda-bersekongkol-permainkan-harga-skuter-matik
Herawati, Yunisa. 2018. The fate of the Yamaha-Honda Cartel is in the Hands of the Supreme Court. Accessed on December 25, 2018, from https://www.viva.co.id/otomotif/motor/1011748-nasib-kartel-yamaha-honda-di-tangan-mahkamah-agung
KPPU. 2016. Indonesian republic's business competition supervisory commission. Copy of case decision number: 04/KPPU-I/2016.
KPPU. 2016. Decision on Case No. 04/KPPU-I/2016. Accessed on December 25, 2018, from http://www.kppu.go.id/id/blog/2017/02/putusan-perkara-no-04kppu-i2016/

KPPU. 2019. KPPU and BPKN Protect Consumers from Cartels. Accessed on June 18, 2019. From http://www.kppu.go.id/id/blog/2019/05/kppu-dan-bpkn-lindungi-konsumen-dari-kartel/

KPPU. Guidelines For Article 5 Concerning Price Fixing Under Law Number 5 Year 1999 Concerning Prohibition of Monopolistic Practices And Unfair Business Competition. Accessed on June 18, 2019.

Laskowska, Magdalena. 2014. Oligopoly - how It Works and Why It Works. SSRN Electronic Journal, January.

Purnama, Rayhand. 2017. Story of 'Victims' of Yamaha and Honda's expensive Skutik Cartel. Accessed on December 25, 2018, from https://www.cnnindonesia.com/teknologi/20171208130944-384-261037/cerita-korban-kartel-skutik-mahal-yamaha-dan-honda

Sanusi. 2017. Yamaha and Honda Decided Guilty about the Cartel by Central Jakarta District Court. Accessed on December 25, 2018, from http://m.tribunnews.com/amp/bisnis/2017/12/05/yamaha-dan-honda-diputuskan-bersalah-soal-kartel-oleh-pn-jakpus

Suhendra. 2017. Luring the Yamaha-Honda in the Case of the Skutik Cartel. Accessed on December 25, 2018, from https://tirto.id/menjerat-honda-yamaha-di-kasus-kartel-skutik-cgRF

Supervision of environmental permits as a juridical instrument for protection environment based on community

Endah Pujiastuti, Dewi Tuti Muryati & Tri Mulyani
Universitas Semarang, Semarang, Indonesia

ABSTRACT: Supervision is one of the tools used by the government to enforce legislation. As a juridical instrument, supervision functions to test the effectiveness of the law. In the field of environmental protection and management, policies have been rolled out at the central and regional levels, which also regulate the supervision of environmental permits. This article reviews the environmental permit supervision that is applied based on community involvement in Indonesia. Monitoring of environmental permits based on community involvement is very necessary to be carried out so that activities that affect the environment, especially those that cause environmental damage and pollution, can be anticipated as early as possible. Community involvement since the process of submitting an environmental permit application is the first step that can be taken in the context of environmental permit supervision.

Keywords: permit, environment, community, supervision

1 INTRODUCTION

The Government of the Republic of Indonesia has issued various policies to provide environmental protection and management. This policy is based on the philosophy that a good and healthy environment is a human right and constitutional right for every Indonesian citizen. Supervision is basically preventive law enforcement. According to Bachrul Amiq, that supervision is an effort made so that everything planned by an organization can be carried out properly, it is known the obstacles in its implementation so that resolution actions can be taken so that the planned objectives can be achieved (Amiq, 2013). Referring to the opinion, supervision is also intended to correct if there is a mis-take, in order to prevent the possibility of irregularities. Well-conducted supervision will prevent violations of existing legal norms.

In order to guarantee the protection and management of a good environment, super-vision is seen as one thing that can contribute greatly to the implementation of government policies. Therefore, a good and proper supervision system is of course very necessary. From a juridical perspective, the regulation of supervision is an interesting thing to study, considering that this regulation is the basis for its implementation in the field. Based on this, this article reviews the environmental permit monitoring system as an instrument for protecting the environment in Indonesia.

2 DISCUSSION

2.1 *Environmental permit*

Environment is a resource for human life, brings benefits and plays an important role in health, economy, and social. Human life will never be separated from the environment. The existence of human life is very dependent on the environment. The environment has provided for free various needs for humans to be able to maintain their lives. The available resources from the environment can be utilized properly so as not to cause interference and damage. The use of the environment that is not done well will have an impact on environmental damage or pollution. To prevent it early so that it does not happen, supervision of activities that lead to environmental damage and pollution is needed.

The Law of the Republic of Indonesia Number 32 of 2009 concerning Protection and Management of the Environment (hereinafter abbreviated as UUPPLH) has deter-mined that every person conducting a business and/or activity must have an environmental permit. This environmental permit is a permit granted to every person conducting a business and/or activity that is obliged to have EIA or UKL-UPL in the context of environmental protection and management as a prerequisite for obtaining a business permit and/or activity. Furthermore, determined in Article 36 UUPPLH affirms that every business and/or activity that is obliged to have EIA or UKL-UPL is required to have an environmental permit. This environmental permit is issued based on environmental feasibility decisions or UKL-UPL recommendations determined by ministers, governors, or regents/mayors in accordance with their authority based on the results of the EIA Assessment Appraisal. Environmental permits must include the requirements contained in the environmental feasibility decision or UKL-UPL recommendation.

Furthermore, in the context of the implementation of legislation, the State Minister of Environment issued a Regulation of the Minister of Environment of the Republic of Indonesia Number 16 of 2012 concerning Guidelines for Preparation of Environmental Documents. The document includes an Environmental Impact Analysis (Amdal) document, an Environmental Management Effort - Environmental Monitoring Effort (UKL-UPL) form, and an Environmental Management and Monitoring (SPPL) Statement. The following year the Government of the Republic of Indonesia then issued a policy related to the environment through the Republic of Indonesia Minister of Environment Regulation Number 2 of 2013 concerning Guidelines for the Implementation of Administrative Sanctions in the Field of Environmental Protection and Management. The policy regulates the application of administrative sanctions in the field of environmental protection and management. The administrative sanctions are based on two important instruments, namely supervision and implementation of administrative sanctions.

2.2 *Supervision of environmental permit based on community*

UUPPLH Chapter XII First Section stipulates that ministers, governors, or re-gents/mayors in accordance with their authorities must supervise compliance with business responsibility and/or activities on the provisions stipulated in legislation in the field of protection and management. living environment. In the implementation of supervision, ministers, governors, or regents/mayors designate environmental supervisors (PPLH) as functional officials on the basis of delegated authority. At present, the super-vision is carried out by the Regional Environmental Supervisory Officer (PPLHD) who is appointed and is responsible directly to the Regent/Mayor. The authority and implementation of supervision carried out by PPLHD is based on reports on the implementation of environmental permits and/or public complaints.

Supervision of environmental permits through reports is carried out by making and submitting an implementation report on the requirements and obligations of the environmental permit periodically every 6 (six) months. Besides reporting, environmental permit holders must provide guarantee funds for the restoration of environmental functions. Supervision carried out by PPLHD other than based on reports can also come from public complaints. People who experience environmental problems (the community exists) can make complaints to PPLHD for follow up.

Along with the dynamics of society, policies related to the environment need to be pursued as much as possible in the context of providing protection for the environment. Supervision in the context of environmental protection and management is an action that must be carried out to monitor and assess the level of compliance with the conduct of businesses and / or activities in carrying out businesses and / or activities that cause environmental impacts in the form of pollution and damage to the environment and natural resources to applicable regulations.

As already stated that supervision related to environmental protection and management is carried out through efforts to control and prevent pollution by various activities with various legal instruments, one of which is environmental permits. This environmental permit is a mandatory instrument for business activities that are obliged to prepare an AMDAL or UKL-UPL document and is a requirement to apply for a business permit based on Government Regulation Number 27 of 2012. Supervision of the implementation of environmental permits carried out directly or indirectly by the supervisory apparatus the environment currently involves the community to know the compliance of the person in charge of the business and/or activities against the regulations in controlling environmental pollution after the person in charge of the business and/or activities obtaining environmental permits.

Supervision of environmental permits starting from the process of issuing environmental permits involving the community (affected communities), in addition to officials/parties who have the authority to assess requirements documents for environmental permit applications (Amdal, UKL-UPL). The involvement of the community can be in the form of suggestions, opinions, or responses that are used as consideration for officials/parties who have the authority to issue environmental permits. The involvement of the community, especially those living around business activities in the context of super-vision, is expected to be carried out proactively through observing the impact of these activities, then reporting to the authorities when conditions that lead to environmental damage or pollution occur. With cooperation carried out between agencies responsible for the environment and the community, it is expected that the protection of the environment can be maximized.

During the operation of an activity based on environmental permits that have been issued, the supervisory activities are carried out through several stages, namely the stage of supervision, supervision phase, supervision report stage, and recommendation stage (action plan) supervision. This action plan can be in the form of guidance, reprimand, or the determination of administrative sanctions.

The types of administrative sanctions that can be imposed on the person in charge of the business and/or activity have violated the laws and regulations and the requirements specified in the environmental permit can be in the form of written warnings, government coercion, freezing of environmental permits, re-vocation of environmental permits, and administrative fines. Administrative sanctions in the form of a written warning are applied to the person in charge of the business and/or activity in the event that the person in charge of the business and/or activity has violated the laws and regulations and the requirements specified in the environmental permit. In this case, improvements can still be made and have not caused negative impacts on the environment. Government coercion is a real action to stop violations and/or recover in its original state. This sanction is given after written reprimand is made. Freezing of environmental permits is a sanction given to the person in charge of the business and/or activity, namely in the form of legal action not to temporarily impose environmental permits and/or protection and management permits, which results in the cessation of a business and/or activity. This freezing can be done indefinitely. Sanctions for revoking environmental permits are imposed on violations, for example, the person in charge of the business and/or activity does not carry out administrative sanctions imposed on the government. The administrative penalty is the imposition of an obligation to pay certain amounts of money to the person in charge of the business and/or activities because of being too late to force the government. The imposition of this fine since the period of implementation of government coercion is not implemented.

The supervision that has been carried out at this time is considered to be less able to provide maximum protection for the environment. Proper supervision policies are needed so that problems related to environmental protection and management can pro-vide benefits for all parties. One thing that needs to be observed is that which is related to community involvement. The involvement of this community in monitoring activities towards the implementation of environmental permits. Or in other words, involvement is done after environmental permits have been obtained. This is indeed in accordance with the provisions of applicable laws and regulations, but it is appropriate to involve the community from the beginning or early when the petition for the environmental permit is submitted. By involving the community (especially the affected community) from an early age, it is expected that information on the plan of an activity or its operations for an activity carried out whether it has environmental permits or which are not yet equipped with environmental permits can be immediately reported to the relevant institutions and followed up in accordance with applicable regulations.

3 CONCLUSION

Supervision of environmental permits as a form of law enforcement efforts needs to be prepared appropriately so that problems related to environmental protection and management can be resolved appropriately, effectively, and efficiently and can also pro-vide benefits for all parties. Implementing supervision in the environmental field is car-ried out by PPLHD who are appointed and responsible directly to the Regent / Mayor based on the authority of the delegation. The supervision is carried out based on reports on the implementation of environmental permits and/or public complaints.

Community involvement in environmental permit monitoring activities is carried out at the stage of implementing environmental permits. In order to provide maximum protection to the environment, community involvement needs to be done early, starting from the stage of the process of applying for an environmental permit, whether in the form of suggestions, opinions, or responses (especially from affected communities) that must be considered by officials / party that has the authority to issue environmental permits.

REFERENCES

Amiq, B. (2013) 'Aspek Hukum Pengawasan Pengelolaan Keuangan Daerah dalam Perspektif Penyelenggaraan Negara yang Bersih', *Yogyakarta: LaksBang PRESSindo*.

Imamulhadi. 2013. Perkembangan Prinsip Strict Liability dan Precautionary dalam Penyelesaian Sengketa Lingkungan Hidup di Pengadilan. Mimbar Hukum. Vol. 25. No. 3. Oktober 2013. p.416–432.

Kusuma Dewi, Dahlia., Alvi Syahrin, Syamsul Arifin, dan Pendastaren Tarigan. 2014. Izin Lingkungan dalam Kaitannya dengan Penegakan Administrasi Lingkungan dan Pidana Lingkungan Berdasarkan UU Nomor 32 Tahun 2009 tentang Perlindungan dan Pengelolaan Lingkungan Hidup (UUPPLH). USU Law Journal. Vol. II. No. 1. Januari 2014. p. 123–138.

Mina, Risno. 2016. Desentralisasi Perlindungan dan Pengelolaan Lingkungan Hidup se-bagai Alternatif Menyelesaikan Permasalahan Lingkungan Hidup. Jurnal Arena Hukum. Vol. 9. No. 2. Agustus 2016. p.149–165.

Maydrawati, Tri Rusti. 2016. Tinjauan Hukum Lingkungan dan Kebijakannya terhadap Perlindungan dan Pengelolaan Keanekaragaman Hayati. Jurnal Perpektif Hukum. Vol. 16. No. 1. Mei 2016. p.18–44.

Nurlinda, Ida. 2016. Kebijakan Pengelolaan Sumber Daya Alam dan Dampaknya terhadap Penegakan Hukum Lingkungan Indonesia. Jurnal Bina Hukum Lingkungan. Vol. 1. No. 1. Oktober 2016. p. 1–9.

Pujiastuti, Endah dan Dewi Tuti Muryati. "Kebijakan Perlindungan dan Pengelolaan Ling-kungan Hidup Pada Kabupaten/Kota di Wilayah Provinsi Jawa Tengah". Prosiding SNPK. Vol 1. September 2016, ISSN. 2540-783X. p. 124–134. Tanjung Pinang: Fakultas Ilmu Sosial dan Ilmu Politik Universitas Raja Ali Haji;

_____, Efi Yulistyowati, dan Doddy Kridasaksana. 2012. "Sanksi Admin-istrasi terhadap Pelanggaran di Bidang Perizinan". Jurnal Dinamika Sosial Budaya. Vol. 14. No. 1. Juni 2012. p.1–20.

Pudyatmiko, Y. Sri. 2009. Perizinan – Problem dan Upaya Pembenahan. Jakarta: Grasindo.

Ridwan HR. 2011. Hukum Administrasi Negara. Edisi Revisi. Jakarta: Radjawali Pers.

Saija, Vica J. E. 2014. Wewenang Pemerintah Daerah dalam Pemberian Izin Lingkungan Hidup. Jurnal Sasi. Vol. 20. No. 1. Januari – Juli 2014. p. 68–80.

Sekretariat Negara, 2009. Undang-Undang Republik Indonesia Nomor 32 Tahun 2009 tentang Perlindungan dan Pengelolaan Lingkungan Hidup.

_____, 2012. Peraturan Pemerintah Republik Indonesia Nomor 27 Tahun 2012 tentang Ijin Lingkungan.

_____, 2012. Peraturan Menteri Negara Lingkungan Hidup Republik Indonesia Nomor 16 Tahun 2012 tentang Pedoman Penyusunan Dokumen Lingkungan Hidup.

_____, 2013. Peraturan Menteri Lingkungan Hidup Republik Indonesia Nomor 2 Tahun 2013 tentang Pedoman Penerapan Sanksi Administratif di Bidang Perlindungan dan Pengelolaan Lingkungan Hidup.

Siombo, Marhaeni Ria. 2014. Tanggung Jawab Pemda Terhadap Kerusakan Lingkungan Hidup Kaitan-nya dengan Kewenangan Perizinan di Bidang Kehutanan dan Per-tambangan. Jurnal Dinamika Hukum. Vol. 14. No. 3. September 1014. p. 394–405.

Sutedi, Adrian. 2009. Hukum Perizinan dalam Sektor Pelayanan Publik. Jakarta: Sinar Grafika.

Sutrisno, Bambang. 2013. Kerancuan Yuridis Kewenangan Perlindungan dan Pengelolaan Lingkungan Hidup dalam Perspektif Otonomi Daerah. DIH, Jurnal Ilmu Hukum. Vol. 9. No. 17. Februari 2013. p.19–24.

Emerging Trends in Psychology, Law, Communication Studies, Culture, Religion, and Literature in the Global Digital Revolution – Setiawan & Rahmawati (eds)
© 2020 Taylor & Francis Group, London, ISBN 978-1-03-224216-3

Malacca Portuguese intangible cultural heritage: An approach to cultural mapping

Bruno Capão Rego & Joana Cerqueira Bastos
Instituto Camões, I.P, Lisbon, Portugal

Dyah N.A. Janie
Universitas Semarang, Semarang, Indonesia

ABSTRACT: For five centuries the Malacca Portuguese community have maintained a unique intangible cultural heritage comprising oral traditions, expressions, arts, social practices, rituals, festivities, and knowledge regarding nature. However, at present, most of the traditional culture is vanishing at an unprecedented rate and is on the brink of being lost.

Worldwide, globalization is triggering a cultural change that trails a loss of indigenous identity. Locally, the absence of interaction between Portugal and Malacca, the depopulation, the land grant, and reclamation issues have been fading sociocultural share and exchange. The term "extinction" sums up the foreseeable fate of the Malay Portuguese community.

It is imperative to set up a structure to protect the immaterial heritage. Cultural mapping can achieve that by interacting, recording, exposing, revitalizing, and promoting traditional knowledge. This includes the insertion of information into databases and the processing of information on computer-based systems like GIS.

This paper introduces the Malacca Portuguese intangible cultural heritage and associated issues, and highlights the importance of mapping this. The goal is to establish a guideline for future studies that will be vital for the preservation of Malacca cultural heritage diversity and economy.

Keywords: Malacca, Malacca Portuguese, intangible cultural heritage, cultural mapping

1 INTRODUCTION

From 1511 to 1641 Malacca was a pivotal stronghold of the Portuguese presence in Southeast Asia. The spice trade was under their control, as were the merchant and missionary activities spanning from Burma to Japan (Jayasuriya, 2008). The Portuguese rule of Malacca lasted only 130 years and was followed by the Dutch, British and Japanese government until the country achieved independence in 1957.

After 378 years without a Portuguese jurisdiction in Malacca, there is still a remarkable community that shares cultural identity aspects and even physical features with Portugal. Most live in Malacca's seaside Perkampugan Portugis, also called the Portuguese Settlement, established further south along the coast in Ujong Pasir. Others inhabit different areas. According to Joseph Santa Maria, a social activist and member of the Portuguese community, few families remain in Praya Lane. Nevertheless, the actual number of community members living in Malacca is unknown and demographic oral statistics fluctuate between 1000 and 2000 people.

The propensity to migration has been evident in the demographic evolution of the group after the independence of Malaysia. The life path of many traces a tendency to emigrate to places like Perth (Australia), Singapore, and even the United Kingdom (Sarkissian, 2000).

Hence the population has been declining in Malacca, but a particular intangible cultural heritage (ICH) still prompts a sense of belonging. For the next mention we refer intangible cultural heritage as ICH.

Distinct domains set a single immaterial cultural expression that is increasingly in danger (defined in the UNESCO 2003 Convention for the Safeguarding of Intangible Cultural Heritage, Article 2, and Subsection 2). These include oral traditions and expressions profoundly linked with Malacca Portuguese Creole, also known as Papiah Kristang, but also performing arts, such as the local dance and music, and social practices, rituals, festivities, and lastly knowledge of nature. Globalization is subsequently assimilating and adapting: positively by propelling the integration of societies and negatively through the loss of cultural identity. The inexistent contact between the community and Portugal, the migratory depopulation, the land grant and reclamation, and other local issues have led to a more significant fading.

Consequently, it is crucial to set up a cultural mapping project that records, stores and processes ICH. This will support the establishment of policies and measures for the safeguarding of cultural knowledge, help in the development of cultural activities and strategies for government and non-government organizations (NGOs), revitalize cultural sharing between the younger and older groups and use data and GIS to shape and make available immaterial heritage.

In this regard, the mapping will require different steps and follow some of the guidelines developed by Sipiriano Nemani (2012) and established in his "Pacific Intangible Cultural Heritage Mapping Toolkit." It will involve preliminary research, local permission and networking, a survey methodology, and ultimately dissemination and sharing of ICH. This model will enwrap not only the majority of Malacca Portuguese who inhabit the Portuguese Settlement, but also few families that still live in different areas of Malacca and whose number is undetermined.

The present paper aims to set up cultural mapping guidelines to develop future research that will be central for the protection and recognition of Malacca Portuguese ICH and social-economic growth. This model can be applied widely among other scattered communities in Asia. It will also play an essential role in starting to raise awareness in international and national institutions that may support many of these people.

2 MALACCA PORTUGUESE ICH

UNESCO defines ICH as "traditions or living expressions inherited from our ancestors and passed on to our descendants, such as oral traditions, performing arts, social practices, rituals, festive events, knowledge, and practices (…)" (UNESCO, 2011b). In this sense, the nature of the Malacca Portuguese legacy is mostly more intangible than tangible and encompasses many of the domains determined by UNESCO.

Oral traditions and expressions include a diversity of spoken forms, like tales, chants, songs, prayers etc., and language acts in and molds the way as they are told and affected. They are fragile through being passed on by word of mouth and, consequently, can quickly lose their strength and change. Their preservation is crucial as they are a crucial part of protecting social-cultural identity, knowledge, and keeping collective memories alive (UNESCO, 2011a).

Luso-Malay oral traditions and expressions are in danger of extinction, mostly because the Malacca Portuguese Creole included in the UNESCO Red Book of Endangered Languages is disappearing. This creolized dialect was developed during the contact between Portuguese speakers, locals, and other outsiders that used to sail to Malacca (Baxter and De Silva, 2004). Nowadays, it can only be articulated by the living Malacca community and a group of descendants who migrated to and settled in different areas of Malaysia, Singapore, and Australia.

Day-by-day, the small number of fluent speakers is declining, and a chart straight decreasing line follows the age reduction. The older generation is aware of the detachment by the younger and claim this is a result of a non-existent education in the language and mixed marriages. However, a study from 1999 stated that families are shifting the process by mixing English and Creole at home and as a result English is prevailing as the dominant language among the younger groups. According to this, 62.9% of the target population present more fluency in English than in Malacca Portuguese and only 46% shows proficiency in speaking the local language (David and Noor, 1999).

A few dictionaries and one manual have been published, and recently a Malacca Portuguese app has been developed by the University of Malaya. These tools are crucial to consolidating and elevating the preservation of the Creole because there is less knowledge of written Creole than spoken.

Performing arts encompass dance, music, theatre, and other expressions. They also include touchable elements as costumes, musical instruments, body decorations, and other forms that cannot be neglected by ICH. All these forms are vulnerable to threat due to abandonment and standardization (UNESCO, 2011a).

The Malacca Portuguese traditional dance stands out as one of the leading performing arts, but its dynamic is different to in the past. Later and contemporaneous Portuguese influences have mixed the secular Branyo with new types of dances, thereby creating a hybrid Cultural Show (Sarkissian, 2005) (Figure 1). Currently, three dance groups still perform on an occasional basis: 1511 O Maliao O Maliao Dance Troupe; Troupa Don Marina; and Troupa de Santa Maria. Pure cultural devotion is what keeps them alive.

Music is the other emblematic performing art that is no longer listened to nor played as it was. According to Gerard De Costa, leader of the 1511 group, the genre Mata Kantiga is now practically nonexistent. Even though the groups still dance to the sound of melodies like Tianika and Maliao and some performers often play the emblematic Jingkli Nona, music is losing their vitality since the iconic musicians Noel Felix, and Joe Lazaroo passed away, and Malacca Portuguese Creole started to fade.

Both performing arts create a fascinating tangible cultural heritage of costumes and musical instruments. The costumes range from sarong and kebayas to women's bright and embroidered dresses and men's dark hats and jackets. The concertina denotes a singularity between the different instruments.

Social practices, rituals, and festivities contain a diverse scope of forms, such as rites, traditional games, seasonal festivities, ceremonies, and gastronomy (UNESCO, 2011a).

Catholicism is the main link, which not only induces rituals and annual festivities but also affects social daily life and activities. Many Luso-Malay Catholics frequently attend mass and annually celebrate four distinct festivities: Intrudu, Easter, Festa San Juang, Festa San Pedro, and Christmas. However, people have abandoned some of the old customs, specifically in Intrudu (Figure 2). Weddings and funerals also include Catholic rites and exclusive such as sweet coconut sago.

Additionally, this scope also includes traditional gastronomy and local games. For the MPKK action committee chairman, Martin Theseira, gastronomy highlights include Curry Devil, a local dish only cooked in and by people from the community. According to Joseph Santa Maria, games are no longer played as in the past, including Coco Pankada, Bola Buraku, Piang, Baza and Sanggola.

Figure 1.   The cultural show.

Figure 2. Intrudu.

Knowledge of nature includes diverse areas such as fauna and flora, local ecological wisdom, and rituals (UNESCO, 2011b).

Fishing activity is one of the main crafts of the Malacca Portuguese, whose precedents are genuinely in the sea. A fishermen's account recorded by Angelina Santa Maria, a local anthropologist, denotes how vital is the sea, its elements, and associated rituals for these people. The sea pulls them in, writes their stories, and provides their main meal. Some rub their heels and do the sign of the cross with sand and ultimately throw it into the water to ensure a safe journey out in the sea. A whistle call the wind that is not blowing and the sail must guide their boat. Ultimately the sea is their home.

Global and local issues put in danger all the ICH shared by the Malacca Portuguese, and it is pivotal to recognize this to prevent the loss. Land reclamation and grant issues are the most significant threat for all the domains. An uncertain future based on agreements may take away the land and the sea from the community of the Portuguese Settlement. People cannot share and maintain many of the cultural identity elements without a space in common.

It is also crucial to prevent the impact of global and national influences, specifically the ascension of English as the most used language. The decline in Malacca Portuguese speakers is subsequently fading the share of spoken forms that, if not recorded and promoted, will disappear.

The migratory depopulation has equally affected the preservation of ICH as some community members have taken cultural identity values away with them. Jointly, the absence of contact between Portugal and Malacca have not helped to maintain and preserve the unique ICH and efforts must be made to set a cultural link between them.

3 CULTURAL MAPPING FRAMEWORK

In order to safeguard the Malacca Portuguese ICH and reduce or even prevent the damage of the current global and local issues, it is essential to set up a cultural mapping that records, stores, and processes ICH. The mapping will follow some of the guidelines established by Sipiriano Nemani in his "Pacific Intangible Cultural Heritage Mapping Toolkit" and therefore apply the following steps: preliminary research, approval and networking, methodology, and dissemination and sharing of ICH. This project will interact with the Malacca Portuguese community who inhabit the Portuguese Settlement and Praya Lane.

The first step will include primary research to find out and analyze studies that share an idea of the cultural mapping site, intangible elements, historical and contemporary knowledge. Although this study approaches four domains and refers to a few relevant studies, other information may be available in media, local archives, and online. For that purpose, intensive preliminary research will be carried out.

The second step will see agreements with the local political committee to authorize the mapping and all the interactions with the local population. These will be pivotal not only to have a point of coordination and get the local support but also to apply legal rules and develop international and national partnerships with organizations. Also, an exchange of ideas to raise awareness will be crucial to helping people understand and agree with the aim.

The methodology applied will interact in many ways with the community. It will encompass photography, video and audio recording of the cultural expressions, record accounts, and include an inquiry that will quantify and qualify the ICH still present in families. Microsoft Access and ArcGIS will subsequently support the insertion, analyses, and processing of the recorded data.

After mapping, the dissemination and sharing of the ICH will comprise the establishment and promotion of safeguarding mechanisms for ICH. Hence this unique heritage that composes the Malacca Portuguese identity will endure through time and awareness will be raised in global and local organizations that may support many of these people.

REFERENCES

Baxter, A. N. and De Silva, P. 2004. *A Dictionary of Kristang:Malacca Creole Portuguese. with an English-Kristang finderlist*. Pacific linguistics, Research School of Pacific and Asian Studies.

David, M. K. and Noor, F. 1999. Language maintenance or language shift in the Portuguese settlement of Malacca in Malaysia, *Migracijske teme*, 154, 465–481.

Jayasuriya, S. d. S. 2008. *The Portuguese in the East. 1ª ed*. London: Tauris Academic Studies.

Nemani, S. 2012. *Pacific Intangible Cultural Heritage Mapping Toolkit*. Secretariat of the Pacific Community SPC.

Sarkissian, M. 2000. *DAlbuquerques children: Performing tradition in Malaysias Portuguese settlement*. University of Chicago Press.

Sarkissian, M. 2005. Being Portuguese in Malacca: the politics of folk culture in Malaysia, *Etnografica*, Centro em Rede de Investigação em Antropologia, 91, 149–170.

UNESCO 2011a. *Intangible Cultural Heritage Domains*. s.l.: UNESCO.

UNESCO 2011b. *What is intangible cultural heritage?* s.l.: UNESCO.

Emerging Trends in Psychology, Law, Communication Studies, Culture, Religion, and Literature in the Global Digital Revolution – Setiawan & Rahmawati (eds)
© 2020 Taylor & Francis Group, London, ISBN 978-1-03-224216-3

Controlling the emotions of children with autism with social stories while at school

M.N. Irawan
Universitas Semarang, Semarang, Indonesia

Tay Kok Wai
Universiti Kebangsaan Malaysia

ABSTRACT: Children with autism often have problems in controlling emotions when feeling uncomfortable. This is due to obstacles in social interaction and communication, as well as limited interest. The consequences of these problems often lead to non-adaptive behaviors that interfere with their school experience. Forms of behavior that appear include scribbling on the table, suddenly going to the front of the class and then writing, and then when reprimanded the item is thrown away. This needs special handling by providing social skills training related to emotional control at school. This research aimed to reduce non-adaptive behavior while in school. Participants are autistic children in grade 3 of elementary school. The research method took the form of an experiment by applying social stories given before entering the class. The results of the graph analysis show a decrease in non-adaptive behavior related to emotional control while at school. Thus, in general social stories can be used to control emotional problems and reduce non-adaptive behavior while at school.

Keywords: autistic children, emotional control problems, non-adaptive behavior, schools, social stories

1   INTRODUCTION

Children with Autism Spectrum Disorder (ASD) have major problems in terms of social interaction, social communication, and limited and repetitive interests (DSM V, 2013). One obstacle that arises is emotional control. Situations where they are unable to interact and socially communicate make ASD children frustrated and cause unstable emotional turmoil that often leads to non-adaptive behavioral problems (Mangunsong, 2009). Not infrequently, this non-adaptive behavior disturbs the surrounding environment (Atwood, 2007). When non-adaptive behavior arises, especially in terms of processing and controlling emotions, these children experience deficient social skills (Burkhardt, 2008). Individuals who experience such problems usually do not understand how socially acceptable behavior is reasonable and positive (Rao, Beidel and Murray, 2008; Cotugno, 2009). Reducing the emergence of non-adaptive behavior, especially that related to processing and controlling emotions, can be done through training about the right behavior and expectations. One such method is to use social stories as a training medium. This media was developed by Carol Gray with the main goal of helping autism sufferers understand the situation so that they can use adaptive behavior such as understanding the situation, feelings and thoughts of others, and the appropriate emotions (Crozier and Tincani, 2007; Foster, 2015).

Social stories are chosen as a medium for social skills training because they can be arranged according to needs that serve as specific guidance in understanding the context of social behavior, emotions, including controlling inappropriate or non-adaptive behavior (Smith,

2001; Ozdemir, 2008; Foster, 2015). In addition, social stories use visual means that are appropriate to the way children learn autism, which can be used to support environmental adaptation, visual and auditory responses, as well as the means for communication (Peeters, 2004; Anggarwal and Prusty, 2015). The use of visual methods in social stories is more appropriate because it presents clear and real images, is more interesting, and easier to understand by all ages and also for individuals of various intellectual levels (Baker, 2001).

According to Gray (Pane *et al.*, 2015), social stories are media in the form of stories that describe situations or behaviors accompanied by questions of understanding, role playing, and also practice. Social stories use several types of descriptive sentences that show information; perspectives that show thoughts, feelings, and behavior to others; affirmatives that show confidence; directives that show the direction of behavioral expectations; control to explain the situation; and cooperations that provide information to help learn how to overcome problems and consequences of exercised behavior (Chan and O'Reilly, 2008; Lal and Ganesan, 2011; Weiss, 2013). Several studies have shown the benefits of using social stories to train the social skills of autistic children, including building on appropriate behavior and reducing inappropriate behavior problems (Crozier and Tincani, 2007; Chan and O'Reilly, 2008; Graetz, Mastropieri and Scruggs, 2009; Adeniyi and Olayinka, 2012; Wright and McCathren, 2012; Pane *et al.*, 2015; McGill, Baker and Busse, 2015). In addition, some studies also show that social stories are useful for improving the social skills of autistic children (Herrin, 2004; Ozdemir, 2010; Karkhaneh *et al.*, 2010; Lal and Ganesan, 2011; Golzari *et al.*, 2015; Gül, 2016; Karal and Wolfe, 2018)

In this study, non-adaptive behavior problems that occur in autistic children while at school are in terms of emotional control. Examples of behaviors that arise are feeling disturbed or uncomfortable, and easily getting angry, slamming the door, throwing things, scribbling on the table, and clapping hands. The result of this behavior is that it disrupts their own concentration for learning and that of classmates. Therefore, the purpose and benefit of the study is to provide an overview of how social stories can reduce the problem of non-adaptive behavior of autistic children, especially in terms of controlling emotions when feeling disturbed or uncomfortable such as irritation, anger, or anxiety.

2 METHODS

2.1 *Research methods*

The method used for the study was a single case experimental design with the ABA model. Phase A was the baseline conducted five times and phase B was the treatment or intervention, carried out nine times then followed by measuring phase A again. The measurement target is the emergence of non-adaptive behavior in the form of being unable to control emotions while in school, measured both in the baseline, treatment, or intervention phases. Follow-up was carried out for evaluating and determining behavior after administration of the treatment with social stories.

The research method is supported by interviews and observations both for the screening stage and evaluation. From this stage it is known that the subject's non-adaptive behavior, which happens when they are experiencing emotional changes because they feel uncomfortable or irritated, is clapping their hands, scribbling on the table, going to the front of the classroom and writing on the board and when reprimaded the item is thrown away. Based on these findings, training skills with the social stories media are given emphasis of the story that teaching social skills about emotional control.

The intervention is implemented by reading, conducting questions and answers to determine understanding, practicing emotional control and providing feedback in the form of rewards, namely drawing or scribbling books. Social dtories are given before the subject enters the class.

2.2 *Participants*

The study participant was D, a 11-year-old third-grade autistic child in private elementary schools in Semarang. The subjects was selected based on suspected screening of non-adaptive behavior problems in the form of emotional control while in school. Another requirement was to be able to read, at least simply. Therefore, intelligence tests were also carried out using the Weschler Intelligent Scale for Children (WISC), which showed total IQ 77 (borderline), Verbal IQ 74 (borderline) and IQ 85 (below average) performance. From the results of the intelligence test, the requirements for simple reading were fulfilled.

2.3 *Data*

Analysis of research data using qualitative analysis of case studies using trend graph analysis that illustrates an increase, decrease, or stability of the behavior.

3 RESULTS

The results of graph analysis show that overall there is a decrease in subject D's non-adaptive behavior while in school. At the beginning of the intervention, subject D showed a non-adaptive behavior that was still quite high. However, after going through several treatment sessions, non-adaptive behavior decreased.

In more detail, during the session subject D was still difficult to hold back. When invited to read social stories, D read and turned pages quickly so their impulses must be controlled. In the next session the subject was invited to discuss non-adaptive behavior through questions of understanding and in return was allowed to scribble on books that had been provided as a means to divert or control their emotions.

After the treatment session was complete, the follow-up results showed that when the teacher reminded D by saying "If you have to learn ...?" then child answered "Sit quietly", while folding their arms and fixing their sitting position. The follow-up results also show that there are other effects in terms of transferring activities or controlling emotions: the subject did not scribble on the table again, but instead scribbled in the book even during the lesson. The results show that social stories can reduce non-adaptive behavior, namely emotional behavior when uncomfortable in school. And the subject can transfer of emotion by scribble the book instead scribbe the desk

Figure 1 shows the trend graph from the study results.

4 DISCUSSION

This study aims to reduce non-adaptive behavior in terms of emotional control while in school. Based on the systematic formulation of social stories developed by Carol Gray (Weiss, 2013) and the results of the study it can be said that social stories have a role in reducing non-adaptive behavior in the form of uncontrolled emotional behavior while at school. Although there are new behavioral findings, namely scribbling in books at the time of learning as emotional shifts, overall they show that non-adaptive behavior decreases, which is also indicated when subjects are calmer while in class.

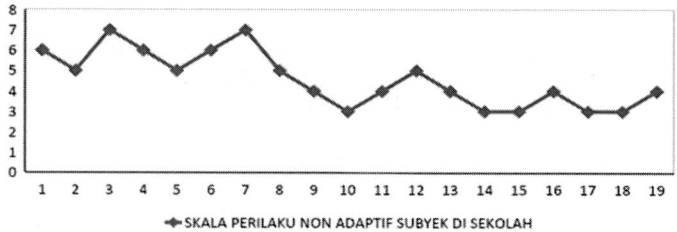

Figure 1. The scale of non-adaptive behavior in schools.

These results are in line with studies that show social stories can reduce non-adaptive behavior in autism, especially behavioral problems that interfere while in school (Adeniyi and Olayinka, 2012). However, from the evaluation of research related to the behavior of scribbling in books as a transfer of emotions, it became another note and evaluation, because for doing it during in the class time.

This is similar to previous research showing that new behavior emerges that is different from the target behavior to be addressed (Irawan and Widyawati, 2018). Therefore, as in other studies, there are several things that need to be considered: the number of participants (Herin, 2004); the involvement of other people, especially parents and follow-up at home (La & Ganesan, 2011; Acar, Tekin-Iftar and Yikmis, 2017); a combination of the use of media or other learning facilities that are accompanied by social stories such as video modeling, reinforcement, prompt, role playing, and so on according to children's needs (Scatonne, 2007; Sansosti and Powell-Smith, 2008; Litras, Moore and Anderson, 2010, Weiss, 2013; Gül, 2016), and criticisms that the tools should be tested for their validity and reliability (Johnson, 2015). Thus, social stories will be more effective if used to train the social skills of autistic children, especially to control emotions and reduce behavior that is not adaptive and prevents new behaviors from arising, as a result of social skills learning outcomes using social stories. Even if new behavior arises, at least the source of the problem is known and can be handled early.

5 CONCLUSION

What can be concluded from this study is that social stories media can control the emotional behavior of autistic children when they feel uncomfortable. That is, in general social skills training with the social stories method can reduce non-adaptive behavior while in school. The study also shows the existence of new behaviors, namely scribbling in books as a transfer of emotions. Therefore, in further research teaching and practicing social skills should be combined with social stories with other forms of intervention models such as video modeling, reinforcement, role play, and or cognitive behavioral therapy. In addition, the number of participants should more than one so they can be compared, and it should involve participants' the parents so that if new behaviors emerge as an effect the intervention can be immediately predicted and handled.

REFERENCES

Acar, C., Tekin-Iftar, E. and Yikmis, A. 2017. Effects of mother-delivered social stories and video modeling in teaching social skills to children with autism spectrum disorders. *The Journal of Special Education*, 504, 215–226.

Adeniyi, B. and Olayinka, M. 2012. Use of Social Stories™ as a Behavioral Intervention for Children with Autism. *Research Paper*. Paper 258. Available at: http://opensiuc.lib.siu.edu/gsrp.

Anggarwal, A. and Prusty, B. 2015. Effect of Social Stories on Social Skills of Children with Autism Spectrum Disorder. *The International Journal of Indian Psychology*, 24.

Atwood, T. 2007. *The Complete Guide to Aspergers Syndrome*. London: Jessica Kingsley Publisher.

Baker, J. 2001. *The social skills picture book*. Arlington, TX: Future Horizons. Inc.

Burkhardt, S. 2008. The challenge of social competence for students with autism spectrum disorders ASD, in: *Autism and developmental disabilities: Current practices and issues*. Bingley: Emerald Group Publishing Limited, pp. 1–24.

Chan, J. M. and OReilly, M. F. 2008. A Social Stories™ intervention package for students with autism in inclusive classroom settings. *Journal of Applied Behavior Analysis*, 413, 405–409.

Cotugno, A. J. 2009. *Group interventions for children with autism spectrum disorders: A focus on social competency and social skills*. London: Jessica Kingsley Publishers.

Crozier, S. and Tincani, M. 2007. Effects of social stories on prosocial behavior of preschool children with autism spectrum disorders. *Journal of Autism and Developmental Disorders*, 379, 1803–1814.

Foster, J. 2015. A Review of the Effectiveness of Social Stories Among Children and Adolescents with Autism Spectrum Disorders. *International Journal of Psycho-Educational Sciences*, 5, 2.

Golzari, F. *et al.* 2015. *Students with Autism Spectrum Disorder*. London: SAGE Publications, pp. 1–8.

Graetz, J. E., Mastropieri, M. A. and Scruggs, T. E. 2009. Decreasing inappropriate behaviors for adolescents with autism spectrum disorders using modified social stories. *Education and Training in Developmental Disabilitie*, 44, 1, 91–104.

Gül, S. O. 2016. The Combined Use of Video Modeling and Social Stories in Teaching Social Skills for Individuals with Intellectual Disability. *Educational Sciences: Theory and Practice*. 161, 83–107.

Herrin, M. J. 2004. Using social stories to teach social and behavioral skills to preschool children with autism. Available at: https://digitalcommons.wku.edu/theses/1112/.

Irawan, M. N. and Widyawati, S. 2018. Penggunaan Social Stories untuk Menurunkan Perilaku Nonadaptif Saat Berada dalam Kendaraan Bagi Autisme Dewasa. *PHILANTHROPY: Journal of Psychology*, 12, 103–114.

Johnson, C. E. 2015. The Effectiveness of Social Stories on Children with Autism Spectrum Disorder: A Literature Review. JMU Scholarly Commons.

Karal, M. A. and Wolfe, P. S. 2018. Social Story Effectiveness on Social Interaction for Students with Autism: A Review of the Literature. *Education and Training in Autism and Developmental Disabilities*, 531, 44–58.

Karkhaneh, M. *et al.* 2010. Social Stories™ to improve social skills in children with autism spectrum disorder: A systematic review. *Autism*,146, 641–662.

Lal, R. and Ganesan, K. 2011. Children with autism spectrum disorders: Social Stories and self management of behaviour. *Journal of Education, Society and Behavioural Science*, 36–48.

Litras, S., Moore, D. W. and Anderson, A. 2010. Using video self-modelled social stories to teach social skills to a young child with autism. *Autism research and treatment*. Hindawi, 2010.

Mangunsong, F. 2009. Psychology and Education Children with Special Needs. Depok: LPSP3 UI.

McGill, R. J., Baker, D. and Busse, R. T. 2015. Social Story™ interventions for decreasing challenging behaviours: a single-case meta-analysis 1995–2012. *Educational Psychology in Practice*, 311, 21–42.

Ozdemir, S. 2008. Using multimedia social stories to increase appropriate social engagement in young children with autism. *Turkish Online Journal of Educational Technology-TOJET*, 73, 80–88.

Ozdemir, S. 2010. Social stories: an intervention technique for children with autism. *Procedia-Social and Behavioral Sciences*, 5, 1827–1830.

Pane, H. M. *et al.* 2015. Evaluating function-based social stories™ with children with autism. *Behavior modification*, 396, 912–931.

Peeters, T. 2004. Autism: Relationship to Theoretical Knowledge and Educational Interventions for Persons with Autism. *Translator: Oscar S. S. Rakyat: Jakarta.*

Rao, P. A., Beidel, D. C. and Murray, M. J. 2008. Social skills interventions for children with Aspergers syndrome or high-functioning autism: A review and recommendations. *Journal of Autism and Developmental Disorders*. Springer, 382, 353–361.

Sansosti, F. J. and Powell-Smith, K. A. 2008. Using computer-presented social stories and video models to increase the social communication skills of children with high-functioning autism spectrum disorders. *Journal of Positive Behavior Interventions*, 103, 162–178.

Smith, C. 2001. Using social stories to enhance behaviour in children with autistic spectrum difficulties. *Educational Psychology in Practice*, 174, 337–345.

Weiss, M. J. 2013. Behaviour Analytic Interventions for Developing Social Skills in Individuals with Autism. *Social Skills and Adaptive Behaviour in Learners with Autism Spectrum Disorders*, 162, 33–51.

Wright, L. A. and McCathren, R. B. 2012. Utilizing social stories to increase prosocial behavior and reduce problem behavior in young children with autism. *Child Development Research*. Hindawi, 2012.

Emerging Trends in Psychology, Law, Communication Studies, Culture, Religion, and Literature in the Global Digital Revolution – Setiawan & Rahmawati (eds)
© 2020 Taylor & Francis Group, London, ISBN 978-1-03-224216-3

Employability factor to improve readiness for changes

Anissa Lestari Kadiyono, Rezki Ashriyana Sulistiobudi
& Muhammad Faris Abdurrohman
Universitas Padjadjaran, Semarang, Indonesia

ABSTRACT: Considering the development of current global economy, every organization is required to continue to adapt and make changes to the progress of their organization. Changes made by the company require the management to prepare their employees to make some changes that are expected to be successfully implemented. This research was conducted to identify what employees need to prepare to deal with changes. Subjects in this research were 92 employees who were fresh graduate from universities in Bandung city which is currently experiencing a change in business direction and implementing new strategies. This research data was obtained through questionnaire method. The questionnaire used in this study was a readiness for changes questionnaire to measure new employee's beliefs, attitudes, and intentions towards changes that occur. Data were analyzed quantitatively through descriptive statistical analysis. The result of this study showed that employees have not reached the condition of complete readiness to deal with changes that occur in the company. It indicates that there are other findings showing that dimensions of beliefs have a significant contribution to employees' readiness for changes. Their employability factor such as their generic skill, emotional intelligence, career development learning, mapping experience of work and life, degree subject knowledge, understanding and skills helped them to facing change and work itself.

Keywords: readiness for changes, employability, beliefs, intentions, attitudes

1 INTRODUCTION

The current global economic development increases competition among companies to become the best one in their fields. This competition requires each company to make organizational changes by adapting and adjusting organizational needs. Organizational change is carried out by the company to maintain its existence and create a better organizational development.

Organizational change is a process of moving from the current condition to conditions expected by an organization in order to increase its effectiveness (Jones, 2013). If a company does not make changes properly, the company will not be able to optimize the effectiveness of its organization. Thus, the company cannot compete optimally in global economic development and will possibly threaten the existence of the organization (Kunert, 2018).

Employability is having a set of skills, knowledge and personal attributes that make a person more likely to secure, and be successful in their chosen occupation (Pool & Sewell, 2007). It is the key for new employee for facing their new job and doing their new task and their new responsibilities. What they do in their job will also facing threatening form external condition, such as organization and economic condition.

The employee readiness for changes is the key to the success of an organizational change made by the company. If employees are not ready to face the changes, they will not be able to follow and will feel burdened with the organizational changes that occur (Hanpachern, 1998). Lack of attention from the company in understanding the processes that occur in employees

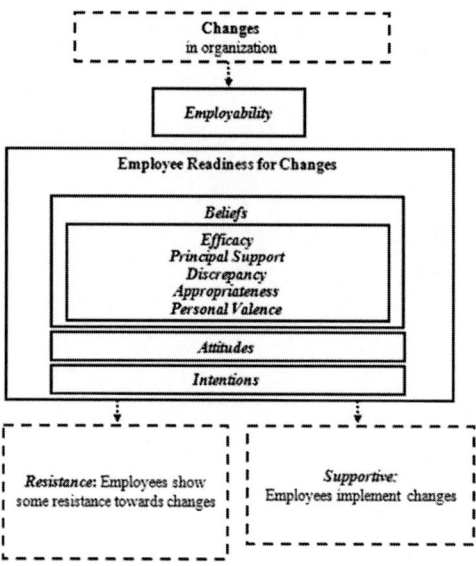

Figure 1. Employee readiness for changes.

causes the company to lose the opportunity to influence employees to be ready to face changes in the company (Rafferty, 2016).

Employee's unpreparedness in the face of changes will bring resistance towards the changes that occur. Resistance occurs because an organization cannot provide an effective change process (Vakola, 2013). Therefore, companies need to increase the employee readiness to make changes.

2   LITERATURE REVIEW

2.1   *Employee readiness for changes*

The concept of employee readiness in the face of change is expressed by Armenakis (1993) as "A cognitive state comprising the beliefs, attitudes, and intentions toward a change effort", which is a cognitive condition consisting of beliefs, attitudes, and intentions towards effort to change.

There are three components forming the employee readiness to deal with changes, namely beliefs, attitudes, and intentions. Beliefs are the evaluative assessment of the whole individual that he is ready to deal with organizational change that is influenced by an individual's belief that change is needed, that he has the ability to succeed in making changes, and that change will have positive results for his work/role (Armenakis, 1993).

Beliefs are defined as an evaluative assessment of the whole individual that he is ready to deal with organizational changes and is influenced by the individual's belief that change is actually needed, that he has the ability to succeed in making changes, and that change will have positive results for his work/role. Armenkis (1993) describes five messages of readiness to change as aspects that describe beliefs, namely efficacy, principal support, discrepancy, appropriateness, and personal valence.

a. Efficacy is defined as "Confidence in individual and group's ability to make the change succeed" (Armenakis, 1993).
b. Principal support is "the extent to which the top leaders, one's immediate manager, and one's respected peers demonstrate that they support the organizational change." Employees' beliefs that emerge from how their teams and supervisors support the changes that occur (Armenakis, 2007).

c. Discrepancy is a "justification for why some change was deemed necessary" or employee's belief that changes are important to do (Armenakis, 2007).

d. Appropriateness is "whether or not the change that is proposed or implemented is the right one for the situation faced by the organization." That is the individual's belief that change will be right for the conditions faced by the organization (Armenakis, 2002).

e. Personal valence refers to the perceived personal benefits (or personal loss) one may reasonably expect as a result of an organizational change" or employee's belief in the equivalent thing he will get based on his efforts towards organizational change (Armenakis, 2007).

Rafferty & Armenkis (2013) proposed definitions of attititudes and intentions. Attitude is cognitive assessment of individuals derived from the various types of information received about change and becomes an overall evaluative assessment of changes in their role (Rafferty & Armenakis, 2013).

While intentions are defined as motivational factors that influence a behavior and are indicators of how hard a person is willing to try and how much effort he or she is willing to exert in order to perform the behavior (Rafferty & Armenakis, 2013).

Intentions are motivational factors that influence behavior and are indicators of how hard someone is willing to try and how much effort he wants to make for the changes faced in his job/role.

The level of readiness will vary for each employee, it is based on the employee's experience of the balance between the costs and benefits estimated by the employee towards the changes that occur (Vakola, 2014). Hence, the condition of employee readiness to deal with changes will affect the employee behavior to support or resist the changes that occur.

3 RESEARCH METHODS

Research design that will be used in this study is a non-experimental with a descriptive approach. The sampling technique that will be used is a non-probability sampling, particularly with a saturation sampling, which is a sampling technique used if all members of the population are used as samples (Sugiyono, 2014). The total number of population in this study was 300 employees. The criteria for this research subject is employees who have worked for at least one year in the company. Therefore, the target population in this study was 92 employees. Data collection will be carried out through questionnaire.

The questionnaire used was the Readiness for Changes questionnaire as the result of the construction of 31 items measuring employee readiness which was adapted from the Readiness to Change theory proposed by Armenakis (1993). This questionnaire has been tested for validity and reliability with the results of reliability of 0.861 as follows:

Dimension	Reliability
Beliefs	0.785
Attitudes	0.740
Intentions	0.720

Subjects were comprised of 67% men, 33% women, 60% married, 38% unmarried, and 2% divorced. A total of 22% have worked for 1 year, 35% worked for 2-5 years, and 43% worked for more than 6 years.

4 RESULTS AND DISCUSSION

Results of this study indicate that the employee readiness for changes was in the high category. The result showed that the cognitive condition of employees was at a high enough level of

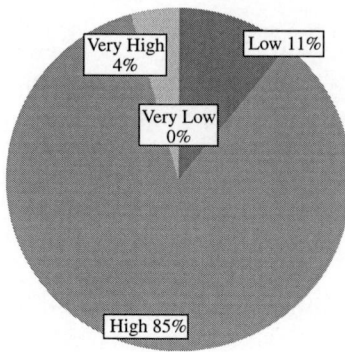

Figure 2.   Employees readiness for changes.

readiness in the face of change. However, there are still some aspects of change that employees are not ready to face, such as changes in the Standard Operating Procedure (SOP), rules, adaptation of government policies, coordination, job description, facilities, and supervisor control.

Standard Operating Procedure; proper understanding of Standard Operating Procedure will improve employee readiness in the face of change. This can increase employee awareness of discrepancy towards SOP applied in the company.

In terms of the rules, employee resistance can arise because employees do not know or understand the new rules set by the company, hence employees still take disciplinary actions from the regulations in the company. Therefore, consistent submission of the applicable rules needs to be carried out by the company.

While in terms of Government Policy Adaptation, companies need to fully convey government policies that have been adapted. This needs to be done because employee resistance can be caused by them not knowing the government policies that have been implemented in the company. Hence, they do not find the benefits of the policy. Submission of information about government policies adapted by the company can improve employee's personal valence.

Viewed from Coordination, with the clear flow of coordination to employees, it will be easier for employees to receive information about what changes need to be prepared.

Job Description; An understanding of employees regarding the standardization of employees in the assignment of responsibilities and tasks for each employee needs to be done by the company. Hence, the efficacy of employees can increase regarding changes in the job description that occur.

Facilities; in order to achieve the expected change goals, the company needs to ensure the availability of facilities that support these changes. When a change is not complemented by the availability of supporting facilities, the employee's efficacy for changes will be affected. Employees are not sure that they will be able to make changes because they will get difficulties from the lack of facilities that support their performance in these changes.

Supervisor Control, where the company needs to set an objective standard of employee assessment. Establishing the same standards of evaluation between employees will disrupt employee confidence that the company will support each employee to make changes fairly. This is able to increase the employee support for changes that occur.

Armenakis (1993) explains that readiness in dealing with changes consists of beliefs, attitudes, and intentions towards the effort to change. Thus, the researcher describes readiness in facing further changes in the dimensions that shape it.

Based on the results of measurements of beliefs, attitudes, and intentions dimensions, it was found that the beliefs dimension has the highest average value among other dimensions. Therefore, it can be concluded that employees have beliefs and preferences to support change, but their willingness to make a change is not in accordance with their beliefs and preferences. This can bring up aspects of change that employees are not prepared to face when implemented in organization.

In further discussion, researchers tried to describe how beliefs, attitudes, and intentions dimensions illustrate employee readiness in dealing with changes in organization. The

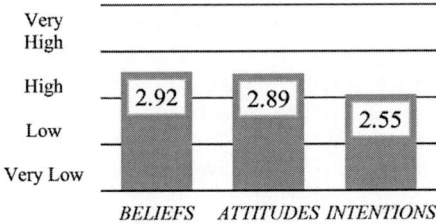

Figure 3. Dimensions of beliefs, attitudes, and intentions.

researchers used a quadratic regression test to analyse the description of beliefs, attitudes, and intentions on readiness for changes.

Based on the results of statistical analysis, it was found that the three dimensions can illustrate the readiness to deal with changes by 53.8%, while the other 46.2% was described by other factors. In addition, it was found that beliefs are the dimension that can best describe the readiness to deal with change (Sig.> 0.05).

Employee beliefs showed a high tendency towards changes that occur in organization. In other words, employees have confidence that change is needed and must be fulfilled, namely in the gap between the desired results and current conditions and changes can be made by themselves and their groups.

Table 2. Model of regression readiness for changes.

Model	R Square
1	,538

Table 3. Regression coefficient of readiness for changes.

Model	t	Sig.
Beliefs	5,684	.000
Attitudes	,508	.612
Intentions	1,968	.052

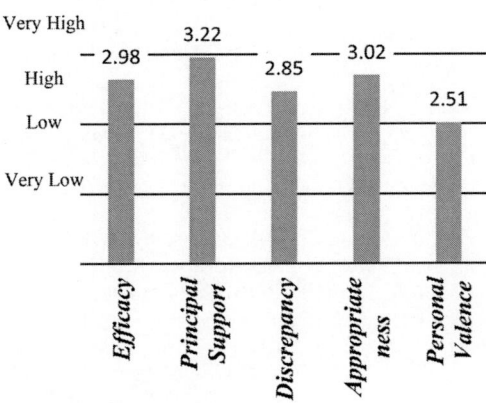

Figure 4. Beliefs sub-dimensions.

These results are based on the table above which shows that efficacy, principal support, discrepancy, appropriateness, and personal valence conditions of employees with a high level as well.

Employees who are ready to deal with changes are those who already have cognitive conditions who believe, are sensitive, and are willing to make changes. This means that employees need to have high beliefs, attitudes, and intentions for the changes that occur. With the presence of high beliefs, attitudes, and intentions in employees, a condition of change that is expected by the organization will be realized.

In this study, it was found that the majority of employees already had readiness in the face of changes. However, there were still a few aspects of change that are still not ready to be faced by them. This is shown from 4% of employees who were in very high category and 85% of employees in the high category.

Meanwhile, there are still few employees (11%) who do not believe in, give preference, and are willing to make an effort to make changes. This is an indication of the unpreparedness of some employees in dealing with the changes that occur. Therefore, the company needs to take action to increase employee readiness in order that the company can realize the expected conditions of the changes that occur.

5 CONCLUSION

The majority of employees have a high readiness in the face of changes. Difference in degrees of beliefs, attitudes, and intentions shape employee readiness and unpreparedness in the face of changes. Beliefs are the variable that best describes the readiness to deal with changes. Employees with managerial positions have higher readiness than employees in staff positions and staff support. The emergence of employee resistance to change can be constrained by supervisor control, facilities, work time, job description, coordination, adaptation of government policies, rules or SOP aspects that need to be reviewed before being implemented in the company according to the employees.

REFERENCES

Armenakis, A. A., & Harris, S. G. (2002) 'Crafting a Change Message To Create Transformational Readines',. *Journal of Organizational Change Management*, 15(2), pp. 169–182. doi:10.1108/09534810210423080

Armenakis, A. A., Harris, S. G., & Mossholder, K. W. (1993) 'Creating Readiness for Organizational Change', *Human Relations*, 46(6), pp. 681–703. doi:10.1177/001872679304600601

Armenakis, A. A., Harris, S. G., Cole, M. S., Fillmer, J. L., & Self, D. R. (2007) 'A Top Management Team's Reactions To Organizational Transformation: The Diagnostic Benefits of Five Key Change Sentiments', *Journal of Change Management*, 7(3-4), pp. 233–290. doi:10.1080/14697010701771014

By, R. T., Armenakis, A. A., & Burnes, B. (2015) 'Organizational Change: A Focus on Ethical Cultures and Mindfulness', *Journal of Change Management*, 15(1), pp. 1–7. doi:10.1080/14697017.2015.1009720

Dacre Pool, L., & Sewell, P. (2007) 'The key to employability: developing a practical model of graduate employability', *Education+ Training*, 49(4), pp. 277–289.

Hanpachern, C., & Morgan, G. A. (1998) 'An Extension of the Theory of Margin: A Framework for Assessing Readiness for Change', *Human Resource Development Quarterly*, 9(4), pp. 339–350. doi:10.1002/hrdq.3920090405

Jones, G. R. (2013) 'Organizational Theory, Design, and Change: 7th Global edition', *Harlow, Essex, England*: Pearson.

Kunert, S. (2018) 'Strategies in Failure Management: Scientific Insights, Case Studies and Tools. Cham', *Switzerland*: *Springer International Publishing AG*. doi:10.1007/978-3-319-72757-8

Madsen, S. R., Miller, D., & John, C. R. (2005) 'Readiness for organizational Change: Do Organizational Commitment and Social Relationships In the Workplace Make a Difference?', *Human Resource Development Quarterly*, 16(2), pp. 213–233. doi:10.1002/hrdq.1134

Rafferty, A. E., & Restubog, S. L. (2016) 'Why Do Employees' Perceptions of Their Organization's Change History Matter?: The Role Of Change Appraisals', *Human Resource Management*, pp. 1–18. doi:10.1002/hrm.21782

Rafferty, A. E., Jimmieson, N. L., & Armenakis, A. A. (2013) 'Change Readiness: A Multilevel Review', *Journal of Management*, 39 (1), pp. 110–135. doi:10.1177/0149206312457417

Sugiyono. (2014) 'Metode Penelitian Pendidikan Pendekatan Kuantitatif, Kualitatif Dan R&D', *Bandung, Indonesia*: *Alfabeta*.

Vakola, M. (2013) 'Multilevel Readiness to Organizational Change: A Conceptual Approach', *Journal of Change Management*, 13(1), pp. 96–109. doi:10.1080/14697017.2013.768436

Vakola, M. (2014) 'What's In There for Me? Individual Readiness To Change and The Perceived Impact of Organizational Change', *Leadership & Organization Development Journal*, 35(3), pp. 195–209. doi:10.1108/LODJ-05-2012-0064

Emerging Trends in Psychology, Law, Communication Studies, Culture, Religion, and Literature in the Global Digital Revolution – Setiawan & Rahmawati (eds)
© 2020 Taylor & Francis Group, London, ISBN 978-1-03-224216-3

The challenges of the online journalism in the industrial revolution 4.0 era in Indonesia

A.P. Lestari & S. Solikhati
Universitas Islam Negeri (UIN) Walisongo, Semarang, Indonesia

Y.B. Setiawan
Universitas Semarang, Semarang, Indonesia

ABSTRACT: The number of hoax news that are still scattered in various online media in Indonesia and the high level of readability of online news in Indonesia is the urgency of why this research is conducted. Given the impact of this industrial revolution 4.0, online journalism is required to be able to present the actual and reliable news. Our research was conducted at suaramerdeka.com, which is part of the printed media business called Suara Merdeka. Suara Merdeka was once appreciated as the best newspaper in Java and Bali as the first best Indonesian regional newspaper. The type of research we use is the qualitative descriptive of critical perspective with Critical Discourse Analysis design. The results of this study refer to the concept of Gatekeeping. As an online media, suaramerdeka.com is closely related to the quantity of news. This system has weakness when pursuing quantity without paying attention to the quality.

Keywords: online journalism, industrial revolution 4.0, Indonesia

1 INTRODUCTION

The industrial world has entered an era called the Industrial Revolution 4.0. This era affects various fields, especially journalism. This change certainly must be followed by the readiness of the journalists to present the actual and reliable news.

Klaus Schwab in his book entitled 'The Fourth Industrial Revolution' (Schwab, 2016) explains that the fourth generation of industrial revolution is characterized by the emergence of supercomputers, smart robots, vehicles without drivers, genetic editing and neurotechnology developments that enable humans to optimize the brain functions. This revolution succeeded in raising the economy dramatically for two centuries after the Industrial Revolution. The indicator was an increase in the six-fold average per capita income of countries in the world.

Next, the second generation of industrial revolution was marked by the emergence of electric power plants and internal combustion motors. This discovery triggered the emergence of telephones, cars, and airplanes that significantly changed the world. Furthermore, the third generation of industrial revolution was marked by the emergence of digital technology.

In the fourth generation of industrial revolution that we are currently facing, we have discovered a new pattern when disruptive technology presents so quickly and threatens the existence of incumbent companies. In this fourth generation of industrial era, the company's agility is the key to success in achieving the achievements quickly.

The effects of the Industrial Revolution 4.0 driven by the Internet of Things (IoT) online strengthen the presence of online journalism. This era is referred to as Digital/Online Journalism, Multimedia Journalism, Interactive Journalism or New Media.

The strengthening of online journalism in the industrial era 4.0 is certainly a tough challenge. The research results conducted by Jon Funabiki and Nancy Yoshihara stated that one thing that should be taken into account so that online journalism persists is increasing the income of operational funding sources (economic) (Funabiki & Yoshihara, 2018).

In addition to challenges in terms of the economy, another major challenges faced by online journalism in the era of industrial revolution 4.0 are the challenges in terms of human resources. The indicator is the number of hoax news aired by online news. The research conducted by Dendy Suseno Adhiarso, Prahastiwi Utari and Sri Hastjarjo (Adhiarso, Utari, & Hastjarjo, 2018) states that the construction of news and responses of netizens have positive and significant influence on hoax news in online media. It means, these two variables have significant influence on hoax news in online media. This proves that the improvement of communication technology has not been followed by an increase in human resource capabilities.

Whereas in Indonesia, online news readers have increased. The Nielsen Consumer & Media View survey until the third quarter of 2017 stated that the reading habits of Indonesians had shifted. The assumption that the media must be free hoists the penetration level of digital media up to 11% with a total of 6 million readers, far more than the printed media readers of 4.5 million people. In fact, the number of printed media readers in 2013 could reach 9.5 million people. Meanwhile, the number of printed and digital media readers approximately is only 1.1 million (Newman, Fletcher, Kalogeropoulos, Levy, & Kleis, 2018).

The technological revolution on the field of rapid information and communication has provided great opportunity for new media to disseminate information to everyone without any limits. The development occurs along with the increasing ease of internet access. The internet has become a necessity for people of digital age (Lestari & Sunarto, 2018).

The number of hoax news that are still scattered in various online media in Indonesia and the high level of readability of online news in Indonesia is the urgency of why this research is conducted. In the era of industrial revolution 4.0, everything is fast. This is a particular challenge for online journalism. Revolution is a very rapid change, but the impact is very vast. Given the vast impact of this industrial revolution 4.0, online journalism is required to be able to present the actual and reliable news. If this is not done, the reader will tend to believe hoaxes that can divide the nation.

Our research was conducted at suaramerdeka.com, which is part of the printed media business called Suara Merdeka Daily Newspaper. Suara Merdeka was once appreciated as the best newspaper in Java and Bali as the first best Indonesian regional newspaper by Nielsen Media Research (Sunarto & Nugroho, 2016). Besides, based on the statement of the Editor in Chief of suaramerdeka.com, Setiawan Hendra Kelana, it is known that suaramerdeka.com is an online newspaper from the printed newspaper-based area that first exists in Indonesia.

The purposes of this research are to describe the professionalism of online journalism that should be done at suaramerdeka.com, to describe the practice of online journalism report arsuaramerdeka.com, and to describe the ability of journalists as the human resources for online news seekers at suaramerdeka.com.

The research on industrial revolution has previously been carried out by Raymond R. Tjandrawinata (2016) who stated that new technology and approaches that combine the physical, digital, and biological worlds fundamentally will change people. Meanwhile, compared to the previous researches on the industrial revolution 4.0, our research has a novelty. The novelty of this research is to answer specifically the challenges of online journalism in terms of Human Resources.

2 RESEARCH METHOD

The type of research we use is the qualitative descriptive of critical perspective with Critical Discourse Analysis design. The study on discourse was also done at the meso level or Discourse Practice analysis. The analysis at this stage is the dimension related to the text/news production process. A news basically is produced through the production process of media texts, such as how the work patterns, the work charts and the routines in producing news (Fairclough, 2010).

Thus, the types and sources of primary data in this study were in-depth interviews with the Editor in Chief and Managing Editor as the controllers of the news production at suaramerdeka.com. While the types and the sources of the secondary data of this study were obtained from: (1) the observations of the worker interactions in daily situations at suaramerdeka.com for one week (February 26, 2018 - March 4, 2018); (2) the written and visual documentation of the results of interactions among the actors at suaramerdeka.com; and (3) the written and visual documentation related to the organizational system of suaramerdeka.com.

3 RESULT AND DISCUSSION

The results of this study refer to the concept of Gatekeeping. The term Gatekeeping is used as a metaphor to describe the selection process in the media work to broadcast news through the media 'door' to the digital platform of suaramerdeka.com.

The idea of Gatekeeping theory is applied into decisions regarding the distribution and the marketing of the news products. In a broader sense, this idea refers to the power to give or to limit access to different voices in the society and often becomes a place of conflict (McQuail, 2011).

The gatekeeper or the keeper of goal who is in charge of making decisions regarding the broadcasting of suaramerdeka.com news products is the editor, Editor in Chief, and Managing Editor. If it is described, the process of gatekeeping at Suaramerdeka.com is as follows.

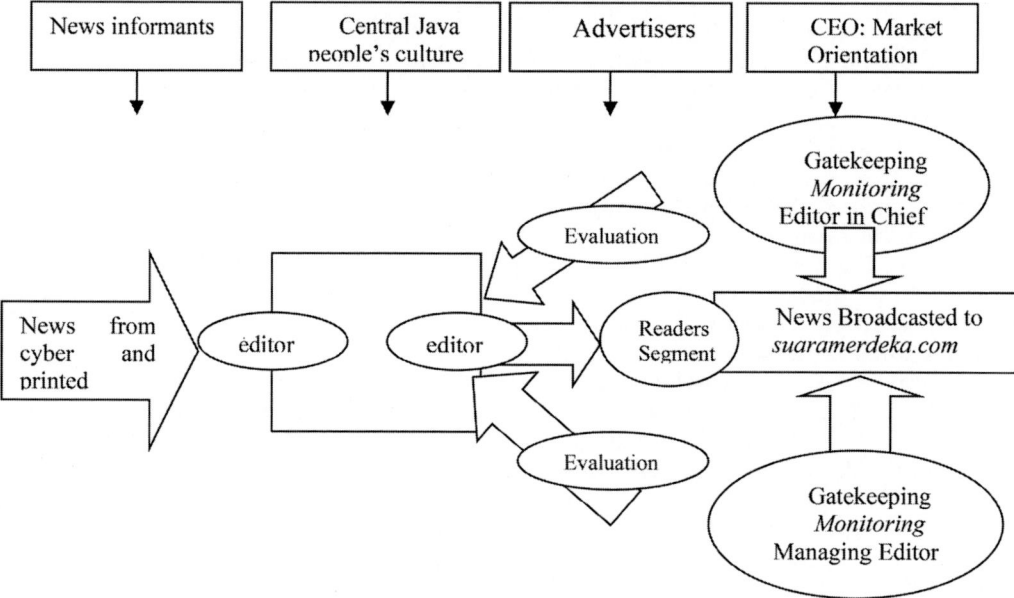

Picture 1.   Suaramerdeka.com gatekeeping model.

As an online media, suaramerdeka.com is closely related to the quantity of news. Chief Editor of suaramerdeka.com targets in 24 hours at least they broadcast 200 news on Suaramerdeka.com site with 8 editors who handle the editing and the broadcasting. The policy is done because if the news aired is less than 200, it cannot be detected by the Google search engine. In addition, the Editor in Chief also plans (not yet realized) to increase the quantity of news broadcasts up to 250 to 300 news each day by adding of the editor.

Furthermore, the Editor in Chief explained that in order to be read by Google, the writing of the title and the news terrace (lead) must also be made different from other online media news narratives that have aired first. This is done because the Editor in Chief hopes that Suaramerdeka.com is in the first rank of search engines and in the future he wants Suaramerdeka.com to become the reference newspaper in Central Java for the readers who want to know the latest news.

Regarding the quantity of news, the editor stated that on average the news from cyber reporters would be published because in 24 hours a lot of news were needed to be aired. When examined, of course this system has weakness when pursuing quantity without paying attention to the quality.

Related to these weaknesses, the Editor in Chief explained that the problem of news searching for online media in the field was due to two factors, namely from the company itself on how to develop strong Human Resources and from Human Resources themselves about how they tried to continuously improve themselves.

One way that Suaramerdeka.com has done to develop Human Resources is by facilitating its journalists to take part in journalist certification organized by the Press Council. The Chief Editor stated that four of the five Suaramerdeka.com journalists had been certified.

Whereas from the factor of Human Resources, the Editor in Chief explained that today's journalists ability have to be improved because of the effects of technological advances so that journalists ignore the principles of cover both side in covering the events.

Evaluating the current journalistic conditions, the Managing Editor explained that when talking about the idealism of journalists with the current technological conditions that were developing and adopted by journalists, made them could not collide. Regarding the current journalist's code of ethics, he stated that code of ethics were sometimes abandoned. However, he added that we could not deny the current journalistic changes.

4 CONCLUSION

Although online journalism pursues actuality, however the principle of accuracy and the balance of news cannot be ignored. This is stated in the Violation of Cyber Media Reporting Guidelines Number 2 regarding the verification and the balance of news, explained in point (b) that news which can harm other parties requires verification on the same news to fulfill the principle of accuracy and balance.

Therefore, training is needed routinely on online media reporters to produce news quickly and accurately. Accuracy and actuality are inevitable in the competition of the journalism industry in the era of industrial revolution 4.0. Accuracy and actuality can be realized by applying communication ethics.

The application of communication ethics is needed because first, the media has powerful power and effects on the public. Second, communication ethics is an effort to maintain a balance between freedom of expression and responsibility. Third, avoiding the negative impact of instrumental logic which tends to ignore values and meaning. Communication ethics want to correct so that these two determinisms are not made as the alibi of the journalists and editors responsibility to justify their mistakes or interests (Libois in (Haryatmoko, 2007)).

The application of communication ethics in online journalism is expected to prevent media organizations from falling in the vortex of market reality. This market reality illustrates how the mass media is under the pressure of tough and tight competitive economy. Competition law requires the mass media to display the latest information, not preceded by other media (Dahlan, 2011). Because this study focused on gatekeeping process on how online news was produced include the capability of those human resources on newsroom, furthermore, recommendation for the next research is about conduct news consumption from the audience to extent the meso analysis level.

BIBLIOGRAPHY

Adhiarso, D. S., Utari, P., & Hastjarjo, S. (2018) 'The Influence of News Construction and Netizen Response to the Hoax News in Online Media', *Jurnal The Messenger*, 10(2), pp. 162–173. https://doi.org/http://dx.doi.org/10.26623/themessenger.v10i2.782

Dahlan, A. C. (2011) 'Hukum, Profesi Jurnalistik, dan Etika Media Massa', *Jurnal Hukum*, XXV (1), 400.

Fairclough, N. (2010) 'Critical Discourse Analysis, The Critical Study of Language', *Malaysia: Pearson Education*.

Funabiki, J., & Yoshihara, N. (2018) 'Online Journalism Enterprises: From Startup to Sustainability', *Renaissance Journalism Center*.

Haryatmoko. (2007) 'Etika Komunikasi: Manipulasi Media, Kekerasan dan Pornografi', *Yogyakarta: Kanisius*.

Lestari, A. P., & Sunarto. (2018) 'Digital Gender Gap for Housewives', *Jurnal The Messenger*, 10(1), pp. 63–71. https://doi.org/http://dx.doi.org/10.26623/themessenger.v10i1.729

McQuail, D. (2011) 'Teori Komunikasi Massa McQuail (terj.) (Edisi 6)', *Jakarta: Salemba Humanika*.

Newman, N., Fletcher, R., Kalogeropoulos, A., Levy, D. A. L., & Kleis, R. (2018) 'Nielsen Reuters Institute Digital News Report 2018'.

Schwab, K. (2016) 'The Fourth Industrial Revolution', *Switzerland: World Economic Forum*.

Sunarto, & Nugroho, A. (2016) 'Laporan Hasil Penelitian: Esensi Pengalaman Kepemimpinan Manajemen Media Lokal di Pulau Jawa. Semarang'.

Responding to the circulation of hoaxes using media literacy and information culture

Tri Mulyani, Dewi Tuti Muryati & Endah Pujiastuti
Universitas Semarang, Semarang, Indonesia

ABSTRACT: This article aims at developing a media and information literacy culture capable of dealing with the increasing daily rate of hoax news. The methodology adopted is normative juridical, with research specifications and descriptive analysis, while the data source consists of secondary and qualitative analysis. The results of the study indicate that literacy culture is a habit adopted to always check the truth of information. This uses information literacy elements such as visual, media, computer, digital, and network literacies, with the ability to understand, analyse and deconstruct media imagery, in order to respond to the circulation of hoax news.

Keywords: hoax, culture, literacy, media, information

1   INTRODUCTION

Technology and technological devices are rapidly developing. According to research, we have been connected to the internet for the past five years in 2018 reached 4 billion, higher than in 2014 which was only 2,4 billion (Hootsuite We Are Social, 2018). This figure shows that the internet penetration rate has reached 52.96% of the total world population of 7.59 billion people, as compared to 35% in 2014. (Dkatadata.co.id, 2018). The internet is used for access to social media (80%), news (75%), entertainment (45%), school assignments (35%), email (25%), financial (10%), buying and selling (5%), and other tasks (5%) (Dkatadata.co.id, 2018). Based on these data, the largest percentage of internet consumers are social media users. Features such as shares, likes, hashtags, and trending topics on social media influences users' interest in reading and consuming public information.

These features enable widespread dissemination of news and information within a short time. This has a positive impact on communities as a target for providing information, and it is undoubtedly a key business area for growing the media industry. Furthermore, the positive impact associated with this growth needs to be balanced with intelligence in processing information, in order not to get distracted by false or fake news (hoaxes) engineered to cover up real information (Indonesia Educate, 2016). Hoax information is aimed at making people feel insecure, uncomfortable, and confused, thereby leading those with poor media literacy skills in the community to make wrong and inconclusive decisions (Indonesia Educate, 2016), so that sanctions can be ensnared regulated by Law Number 19 of 2016 concerning Amendments to Law Number 11 Year 2008 on Information and Electronic Transactions.

This article presents the culture of media literacy and information and public awareness, to develop a community with the ability to understand, analyse and deconstruct media imagery and respond to the circulation of hoax news.

2 DISCUSSION

2.1 *News hoaxes*

Hoaxes are information engineered to cover up real facts using unverified and convincing information. A hoax is also defined as an act of obscuring the facts, by flooding the media with wrong messages to cover the right ones. The purpose is to intentionally make people feel insecure, uncomfortable, and confused, and in response to hoaxes people tend to make weak, inconclusive, and wrong decisions. The development of hoaxes on social media was originally carried out to means to disturb (Indonesia Educate, 2016).

A researcher named Ismail Fahmi (2017) used an empire drone, a software engine created to map the spread of hoaxes on social media. According to the study, 92.40% of hoaxes in Indonesia are spread through social media such as Facebook, Twitter, Instagram, and Path, 62.80% are spread through chat applications such as WhatsApp, Line, Telegram, while 34.90% are spread through websites. Regarding hoax format, 62.10% are spread in the form of writing, while 37.50% are two-dimensional images. Fahmi's research found that the most popular hoax is about a socio-political issue (91.80%), which explicitly discusses elections and government policies or performance. This is followed by hoaxes about SARA (Suku Agama Ras Antar-golongan) at 88.60%, and health issue.

2.2 *Getting to know media literacy and information*

Every day, people are treated to numerous and diverse information, and thereby the use of literacy is important. According to the State University of New York, information literacy is the ability to recognize when information is needed, placed, evaluated and effectively/simultaneously utilized in communicating (Ainiyah, 2017). Literacy is the ability to read and write, using information from various forms. However, besides reading and writing, there are five types of literacy that act as information elements:

1. Visual literacy: the ability to understand and use images to think, learn and express oneself.
2. Media literacy: the ability of citizens to access, analyse and produce information for specific results according to the National Leadership Conference.
3. Computer literacy: the ability to create and manipulate documents and data using word processing software, databases, etc.
4. Digital literacy: a skill related to mastering digital resources and devices.
5. Network literacy: the ability to access, place and use information in the world of networking (the internet).

Based on the description, a literate society has the ability to recognise when information is needed, placed, evaluated and used effectively/simultaneously in communicating various forms and types of information.

Furthermore, the media is required to deliver information accurately. The word media is derived from medium, which means intermediary or introduction (Sadiman, 2008). Therefore, anyone able to understand the source, communication technology, the code used, the information produced, the selection of interpretations, and the impact, is said to have media literacy (Rubin, 1998). This is the process of accessing, and critically analysing the information contained in the media, while creating information using the various media tools (Hobbs, 1998). According to (Rahmi, 2017) the purpose of literacy media is to help people develop a better understanding and control media influences in their daily life. Control begins with the ability to see the difference between media information capable of improving a person's quality of life and destructive media information.

Using media literacy and information is essential, especially in this era of technological advances, as mastering this makes people aware of hoaxes.

2.3 *Responding to the circulation of hoaxes using media literacy and information culture*

The circulation of hoax news through social media, according to the author, can be suppressed by developing a culture of media and information literacy. Culture is a way of life that is developed and shared by a group with an inheritance from one generation to another, while literacy is the ability to read and write. According to the Program for International Student Assessment, out of 65 countries surveyed in 2012 the literacy culture in Indonesia ranked in 64th place.

Furthermore, UNESCO stated that Indonesia scored 0.001%, which means that out of 1000 people only one person had an interest in reading. Reasons for this are that the habit of reading and writing had not yet begun at home, sophisticated technological developments, minimal reading facilities, a lack of motivation, and students' lazy attitudes to developing ideas (Ainiyah, 2017). Literacy culture is the tendency of the population to check the accuracy of information using the information literacy elements such as visual, media, computer, digital, and network literacies, which gives people the ability to understand, analyse and deconstruct media imagery, so as to respond to the circulation of hoax news.

3 CONCLUSION

Getting to know media literacy and information for some one in the technology advances this important and many benefits, besides getting a sense of security, comfort an not confusion, and being able to make the right decisions in the midst of the circulation of hoax news that was deliberately created for the benefit of someone, can also avoid law problem.

REFERENCES

Ainiyah, Nur. 2017. Membangun Penguatan Budaya Literasi Media dan Informasi dalam Dunia Pendidikan. Jurnal Pendidikan Islam Indonesia, Volume 2 Nomor 1, Oktober 2017, hal 65–77.

Fahmi, I. 2017. Perilaku Masyarakat Indonesia Terhadap Hoax, Media dan Budaya Baca (https://www.slideshare.net/IsmailFahmi3/perilakumasyarakat-indonesia-terhadap-hoax culturalmediadanread/, diakses 15 April 2019), 2019.

Hobbs, R. dalam Gracia Rahmi Adiarsi, Yolanda Stellarosa, dan Martha Warta Silaban, 1998. Literasi Media Internet di Kalangan Mahasiswa. Jurnal Humaniora Volume 6 Nomor 4 Oktober 2015, hal 473.

Hootsuite We Are Social. 2018. Digital in 2018: World's Internet Users Pass The Billion Mark (https://www.wearesocial.com/us/blog/2018/01/global-digital-report-2018/, diakses 15 April 2019), 2019.

Rahmi, 2013. dalam Gracia Rahmi Adiarsi, Yolanda Stellarosa, dan Martha Warta Silaban, 1998. Literasi Media Internet di Kalangan Mahasiswa. Jurnal Humaniora Volume 6 Nomor 4 Oktober 2015, hal 472.

Rubin, A. 1998. dalam Gracia Rahmi Adiarsi, Yolanda Stellarosa, dan Martha Warta Silaban, 1998. Literasi Media Internet di Kalangan Mahasiswa. Jurnal Humaniora Volume 6 Nomor 4 Oktober 2015, hal 473.

Sadiman, A. (2008) Media Pendidikan: Pengertian, Pengembangan dan Pemanfaatannya Jakarta: PT. Raja Grafindo Persada.

Sekretariat Negara Republik Indonesia, 2016. Undang-Undang Nomor 19 Tahun 2016 tentang Perubahan Atas Undang-Undang Nomor 11 Tahun 2008 tentang Informasi dan Transaksi Elektronik.

School resilience and religious radicalism in senior high schools

Joan Hesti Gita Purwasih & Ahmad Arif Widianto
Universitas Negeri Malang, Malang, Indonesia

ABSTRACT: Severe radicalism infiltration has spread in senior high schools, which should be a society educating agent. Students are now be exposed to incessant radicalism in schools. This study was conducted to analyze the school resilience system and the radicalism infiltration process among public senior high school students. Using qualitative research and AGIL (adaptation, goal attainment, integration, latency) theory, this study mapped the elements, systems, vulnerability, and immunity of senior high school resilience in dealing with radicalism. The result of research showed that the schools are vulnerable to radical thinking because of the schools' poor resilience. The school elements' (particularly teachers') awareness of the radicalism threat is still very poor. The school bureaucracy has not been well-coordinated thereby becoming the entry point for radical groups. Teachers have different religious perspectives and expressions and they have very limited control over the students' activity out of school.

1   INTRODUCTION

In recent years, Indonesia has faced religious radicalism challenges, threatening the stability of national and state life. Radical actions arise in many elements and social institutions. Religious radicalism is instead growing stronger among the educated population, who should be consciousness of living within the nation and state (Ali and Purwandi, 2018; Saputra, 2018). Recent surveys show a high percentage of radicalism among the educated population, particularly students. The results of a survey conducted by Mata Air Foundation and Alvara Research Center in 2017 showed that 23.3% of senior high school students agree to do jihad and similar data was released by the State Intelligent Agency (BIN). A survey by PPIM UIN Jakarta in 2018 found that 58.5% of students and college students has radical religious perspectives. These results give stringent warnings to prevent the dissemination of student radicalism in the school environment.

On a practical level, the result of our survey of 300 students in 2018 showed that radical thinking among senior high school students in Malang and Batu Cities is becoming a real risk. We found that some students have low trust in Pancasila thought and 2% of college students in Malang and Batu Cities strongly agree that Pancasila can be replaced with new ideology. Another phenomenon leads to the idea of replacing the Indonesian government system with a khalifah system, and students in Malang and Batu Cities strongly agree (8%) and agree (23%) with this.

Some previous studies also showed that the entry of radicalism into the school environment occurs in the following ways: religious extracurricular activities (Salim, Najib Kailani and Azekiyah, 2011, Rokhmad, 2012, Akmaliah, 2013, Wahid Foundation, 2016);school student organizations (Darraz and Qodir, 2018);teachers with radical thoughts (PPIM, 2016); school alumni affiliated with radical groups (Masooda, 2016); text books and student work sheets with radical Islamic material (Rokhmad, 2012, PPIM, 2016, Hasan, 2018); and a radical-indicated education curriculum (Zainiyati, 2016). The findings of the studies above confirm that religious radicalism in the school environment is getting stronger through an infiltrating process and the media, and such aspects as alumni extracurricular activities, teachers, and the headmaster's policy (Maarif Institute, 2018).

This research is different from previous studies, analyzing the radical thinking phenomenon from a school resilience aspect. School resilience is an environment or area for organizing the learning process, a place to inculcate ideological, personal and religious values and a variety of sciences, technologies, and skills. School as the learning process site should protect from any negative effect that can harm the learning process, which may come both from inside and outside school. The school may not be used for purposes other than educational objectives. The objective of education has been mentioned in Article 3 of National Education System Law. Therefore, school resilience is an attempt to prevent, anticipate, and cope with negative deeds committed by parties, either internal or external, that can result in hazards that are criminal in nature, such as illicit drugs, pornography, and violence in the school environment, and that will disturb the school's composure, orderliness and safety. The school resilience perspective as a system and function becomes the author's point of view in explaining the radical thinking phenomenon.

2 METHOD

This research employed a phenomenological qualitative research method. The unit of analysis in consisted of three schools in Malang and Batu Cities. Teachers and students' joint experience with the radicalism phenomenon was explored using FGD and in-depth interviews. FGD was selected as an appropriate method to explore the teachers' joint experiences in treating the radicalism phenomenon. In-depth interviews were also conducted with students who had experience of invitations to radicalism. Data validation was conducted using source triangulation to find specific and rich data. Furthermore, the author applied Miles and Huberman's interactive model of analysis to map the school resilience system, which became the focus of research.

3 RESULT

3.1 *School feels secure from radicalism threats*

Several terroristic bomb actions committed by radical groups occurred in 2018 in Indonesia, which was of great concern to the nation and across the world. In reality, such a social phenomenon has not yet attracted serious attention yet from schools in treating radicalism threats. The data from FGD and teachers in some schools shows that there has been no special program to fortify the schools in relation to radicalism. Some teachers, however, feel that the licensing and procedures to allow outsider groups entry to schools has been sufficient.

Some teachers recounted experiences with the radicalism phenomenon during FGD. Other teachers and school officials, in fact, did not have such experience. There is no space accommodating this discourse phenomenon in schools and therefore some teachers are experiencing and solving this problem alone. Examples of radicalism phenomenon experienced are as follows: 1) students feel antipathy towards strangers due to different religions; 2) students are invited to join religious groups leading them to do jihad; 3) alumni they know teach them to wage war on different religious groups. When Focus Group discussion is held, the teacher do not believe this is occurring and are sure that their school is very secure from radicalism.

3.2 *Teachers' different religious perspectives*

Different perspectives on undertaking worship are reasonable phenomena found in society. Such conditions also occur in the school environment. Teachers have different interpretations of the apparent changes in how students are actualizing worship. For example, some students no longer want to shake hands with those of opposite sex and change their dress corresponding to the religion tenet. To some teachers, this is reasonable; however, to others it is a phenomenon to which we should be alerted.

This condition becomes a sensitive issue for discussion. The reality of different religious beliefs and sects is part of Indonesian culture in general. Therefore, when this issue is discussed in various media, there is a risk it will generate religious sensitivity. Not all teachers are open to and accept discussion of such conditions, and as a result the phenomenon of student's behavioral change is often neglected.

3.3 *Many bureaucratic fissures in school*

Schools have systems to limit the involvement of outsiders. This system can be seen from the schools' rules and the duty of the school apparatus, such as security, public relation divisions, and deputy headmasters. Everyone who wants to visit and to conduct any activity should report first to their corresponding contact. However, in reality this does not always happen and observations show that close relationships with some teachers creates many opportunities for radical groups to enter into the school.

Teachers have the authority to bring other groups into schools and some school activities are held in non-procedural ways. Coworkers' willingness and abilities to address this with such teachers is often seen as taboo. This attitude builds on the reluctance and lack of care about dysfunction of the schools' procedures.

3.4 *Students' activities out of school are not the school's responsibility*

Students' activity out of school is not considered as the school's responsibility, as suggested by teachers during the FGD process. Teachers, as part of the school apparatus, feel they have limitations in such supervision and that children's behavior out of school is the parents' responsibility. Meanwhile, in some cases the influence of radicalism at school successfully leads the children to leave school.

This responsibility limitation becomes an excuse to give the students maximum service performance and teachers' integrity as an agent that supervises, takes care of, and drives school resilience cannot be maximal.

4 CONCLUSION

The phenomenon of radicalism occurs in the school environment. The penetration and infiltration of radicalism tenets at school are clear through various media and processes. However, schools are not really aware of such a threat and this is the main cause of weak school resilience. The author used a functional structural perspective suggested by Talcott Parsons to analyze this phenomenon and also AGIL (adaptation, goal attainment, integration, and latency) theory (Segre, 2012). The existing system is not changing to capture the risk of social change in the school environment and this indicates the failure of the adaptation function, and the fact that schools do not see radicalism as a threat indicates that the goal attainment function is not performing well. Sensitivity in examining ways of worship suggests a weak integration function. Meanwhile, in relation to the latency function, schools have not yet made further attempts to treat the threat of radicalism. The government must launch a special program on antiradicalism in school as follows: first, students, parents, and teachers must helped to understand the dangers of radicalism, for example through education on social media; second, the government can work with universities to develop early detection instruments of radical thinking and design an ideal school defense system.

REFERENCES

Ali, H. and Purwandi, L. (2018) 'Radicalisme Rising Among Educated People', *Alvara Reseach Centre*.
Abuza, Zachary. 2007. *Political Islam and Violence In Indonesia*. New York: Routledge.

Ahmad Gaus AF, 2013. Pemetaan Problem Radikalisme di SMU Negeri di 4 Daerah. *MAARIF: Menghalau Radikalisasi Kaum Muda: Gagasan dan Aksi.* 8, 1.

Al-Qurtuby, Sumanto, 2010. In Search of Socio-Historical Roots of Southeast Asia`s Islamist Terrorism with special Reference to an Indonesian Experience. *Journal of Indonesian Islam* 4, 2.

Azca, Najib. 2013. Yang Muda, yang Radikal: Refleksi Sosiologis terhadap Fenomena Radikalisme Kaum Muda Muslim di Indonesia pasca orde Baru. *JurnalMa`arif* 1, 8.

Bruinessen, Martin Van. Genealogis of Islamic Radicalism in Post-Suharto Indonesia. *South East Asia Research*, 10, 2, 117–154.

Darraz, M. A. and Qodir, Z. (2018) *Osis Mendayung di Antara Dua Karang: Kebijakan Sekolah, Radikalisme Dan Inklusivisme Kebangsaan.* Jakarta: MAARIF Institute for Culture and Humanity.

Gall, Meridith D., Gall, Joyce P, Borg, Walter R. 2003. *Educational Research An Introduction.* USA: Pearson Education.

Hasan, N. (2018) Literatur Keislaman Generasi Milenial: Transmisi, Apropriasi, dan Kontestasi. *Literatur Keislaman Generasi Milenial: Transmisi, Apropriasi, dan Kontestasi*, pp. 1–27.

Masooda, B. (2016) *Study on Islamic Religious Education in Secondary Schools in Indonesia.*

Munip, Abdul. 2012. Menangkal Radikalisme Agama di Sekolah. *Jurnal Pendidikan Islam* 1, 2.

Policy Brief Series. 2018. Potret Keberagaman Kaum Muda Muslim Indonesia.

Policy Brief Series. 2018. Menolak Radikalisme dalam Pendidikan. Mencipta Sekolah Inklusif-Kebinekaan, 7, 1.

Rokhmad, A. 2012. Radikalisme Islam dan Upaya Deradikalisasi Paham Radikal, *Walisongo: Jurnal Penelitian Sosial Keagamaan*, 20, 1, 79–114.

Salim, H., Najib Kailani, N. and Azekiyah. 2011. *Politik Ruang Publik Sekolah Negosiasi dan Resistensi di Sekolah Menengah Umum Negeri di Yogyakarta.* Yogyakarta: CRCS.

Saputra, R. E. 2018. 'Api dalam Sekam: Keberagaman Generasi Z', 1(1).

Segre, S. 2012. *Talcott parsons: An introduction.* University Press of America.

Sisdiknas. 2003. Undang-Undang Sistem Pendidikan Indonesia.

Survei Nasional: Sikap dan Perilaku Keberagaman di Sekolah dan Universitas. 2018. PPIM UIN Jakarta- UNDP Indonesia.

Wahid Foundation. 2016. 'Riset Potensi Radikalisme di Kalangan Aktivis Rohani Islam Sekolah-Sekolah Negeri'.

Wahyudi Akmaliah Muhammad dan Khelmy K. Pribadi. 2013. Anak Muda, Radikalisme dan Budaya Populer. Journal MAARIF Vol. 8, No. 1 Juli.

Widianto, Ahmad Arif, Joan Hesti GP, Nanda Harda Pratama, M. Rani Prita Prabawangi. 2018. Living Pancasila: Deradikalisasi agama melalui implementasi nilai-nilai pancasila di kalangan pelajar di kota Malang dan Batu (2017–2018). Laporan Penelitian. Tidak Dipublikasikan. Universitas Negeri Malang.

Yusar. 2015. Perlawanan Kaum Muda terhadap Hegemoni Radikalisme Agama dalam Bentuk-bentuk Budaya Populer. *Jurnal Ilmu Sosial Mamangan*, 2, 1.

Zainiyati, H. S. 2016. Curriculum, Islamic understanding and radical Islamic movements in Indonesia, *Journal of Indonesian Islam*, 10, 2, 285–308.

_____. 2015. Pertemanan Sebaya Sebagai Arena Pendidikan Deradikalisasi Agama. *Walisongo*, 23, 1.

_____. 2016. The Youth, The Sciences Students, and Religious Radicalism. *Al-Ulum* 16, 2.

Emerging Trends in Psychology, Law, Communication Studies, Culture, Religion, and Literature in the Global Digital Revolution – Setiawan & Rahmawati (eds)
© *2020 Taylor & Francis Group, London, ISBN 978-1-03-224216-3*

Implementation of occupational health and safety management system in human resources development to improve performance

Hardani Widhiastuti, Gusti Yuliasih & Yudi Kurniawan
Universitas Semarang, Semarang, Indonesia

ABSTRACT: This study emphasizes improving the performance of human resources development through the K3 management system. Many situations that can benefit from K3 occur in workplaces. Accidents are reported in various print and online media. It is therefore necessary to study the factors that optimize the performance of human resource development through a good K3 management system to manage stress caused by jobs full of challenges. To date, the better management of K3.Occupational Health and Safety System (SMK3) in Indonesia has not received much attention. PT PLN Persero is one of the states in Indonesia to implement a K3 management system, especially the distribution network. A growing number of accidents still occur, especially in high-risk occupations associated with contractors or electrical voltage, despite improvements and socialization. Our research is an attempt to anticipate accidents and occupational health hazards through a system that already exists in the company and including the associated development of other variables such as performance and human resource development. This research is also a first step in producing the occupational health and safety system (SMK3) model.

Keywords: human resources performance development, occupational health and safety system, SMK3

1 INTRODUCTION

Human resource development has become an important part of the efforts to manage human resources as a whole. In essence, human resource development has a broad dimension that aims to improve employee potential, in an effort to increase professionalism in the organization.

Planned and organized development of human resources with proper management will be able to save other resources, or at least increase the efficiency and effectiveness of processing and usage of organizational resources. Work is an expected aspect of life for most people. No one does it mean in life, such as with their job. By working, people can socialize and communicate with others. At the same time, in performing the work, one can apply the knowledge and skills of the operation.

Many cases where K3 would be beneficial occur in companies, as various reports on workplace accidents have appeared in print and online media (Muchtaruddin, 2014). The BPJS Center reported that 8,900 occupational accidents occurred nation-wide from January to April 2014 (Mohari, 2014), while 2,500 cases occurred in the Riau Archipelago. And, just what happens is in the armed forces, the lack of attention to the system in the work environment, to their production employees who died in the coil of hair employees sign such production.

The purpose of the research is to find a model Safety and Health Management System (SMK3) to reduce the stress of work and increase motivation via development of human resource management following ideal conditions. It is based on input from PLN Distribution

Central Java and Yogyakarta area in general and Field Employee Central Java Province and Yogyakarta PLN in particular in order to devise appropriate strategies related to the effort to improve the performance of human resource management in relation to the Health and Safety System (SMK3).We hope that this research can help encourage human resource management research where the conceptual and measurement models have not been developed well, especially in the relationship between System Occupational Health and Safety (K3)and development of human resource management. We also hope that this research can contribute ideas for the development of knowledge in human resources research of practical importance for managerial practices to improve development of human resource management.

2 LITERATURE REVIEW

2.1 Occupational health and safety system (SMK3)

K3 is very important for everyone working in a corporate environment, especially those engaged in the production. It is especially important for understanding of health and safety at work on a daily basis to improve performance and prevent potential losses by the company. However, the question is the extent to which the company is obliged to follow the principles of K3 in its work environment.

Kani et al. (2013) found that many workers do not know about K3— its meaning K3, how it is implemented, and so forth. This shows that there is still a lack of attention or commitment of the contractor to carry out the K3 program well, which ultimately has an impact on the effectiveness of human resource development.

We can conclude that the government should make rules for the implementation of K3 to impose safety requirements so that the potential hazards of workplace accidents can be eliminated.

Law No. 1 of 1970 has the goal of keeping the public and the working environment safe, healthy, and prosperous, which in turn will improve all-around productivity efficiency. The main provision of the Act is a system of prevention and implementing K3 within a business unit, with its rights, obligations, responsibilities, and sanctions as well as development work. Regulations on SMK3 were refurbished with Government Regulations No. 70 of 2012 (Peraturan Perundang-Undangan Nomor 70 tahun 2012, 2012).

2.2 Safety and health management system (SMK3) motivation management

In this study, the emphasis has been more on motivation according to the Achievement Motivation Theory (Achievement Motivation) of McClelland. The theory states that an employee has potential energy that can be utilized depending on encouragement, circumstances, and opportunities. To promote workers' morale there is a need to provide encouragement to excel, excel in connection with a set of standards, and struggle to succeed. This includes the need for power: the need to get people to behave in a way that they would not have if not forced to do so.

This also includes the need for affiliation: the desire for interpersonal relationships that are friendly and intimate (16). Books (18), revealed that SMK3 will significantly increase motivation. When SMK3 is successfully carried out within the company, it will have a positive influence on employee motivation. An empirical relationship between SMK3 and motivation can be explained in the research done by (4). Based on his research, Mahruzar found a positive and significant relationship between providing safety assurance at work and work motivation. This suggests that with higher levels of safety assurance employee motivation is also higher, and conversely the lower the level of safety assurance, the lower the work motivation. The effective contribution of guaranteed safety to work motivation is 94.7%. Hence we propose the following hypothesis:

H1: There is an influence of the Safety and Health Management System on employee motivation; in this case it is the motivation contributed by the K3 management.

2.3 *Work stress*

Robbins and Mary (2002) suggest that stress is a dynamic condition in which there is an opportunity for confrontation in order to a person's desires to be perceived. Handoko, consider stress to be a condition that affects the emotions, thoughts, and the person's condition.

Robbins and Mary (2002) describe stress more on the basis of its physiological symptoms, namely that stress can create a change in metabolism, increased heart rate and breathing rate, increased blood pressure, headaches, and heart attack. Psychological symptoms include depression, anxiety, tension, anxiety, irritability, and procrastination. Behavioral symptoms, such as decreased productivity, absenteeism, and increased turnover (20) and (13) caused by stress in the work environment can affect motivation. The research results illustrate that good stress management will have an impact on increasing the motivation of employees. Thus the following hypothesis is proposed:

H2: There is an influence of stress on the employee's job motivation.

2.4 *Motivation management*

According to Robbins and Mary (2002), motivation is a process that explains the intensity, direction, and persistence of efforts to achieve purposes. Motivation is a reaction that begins with the need to create a desire or goal, further raises tension (strain), namely desires that have not been fulfilled, which leads to an onset of action that leads to the goal eventually being satisfied. In the organizational behavior literature, various research indicates that motivation has a positive influence on performance. Classical studies (14) showed a positive and significant relationship between motivation and performance. Thus, the research examines the relationship between performances in the context SMK3 motivation that so far has quite rarely been conducted. According to Hendarman, motivation is a factor that has a positive effect on performance. Thus the hypothesis is proposed:

H3: There is an influence of motivation, in this case that generated by K3 management, on the performance of human resource development.

2.5 *Performance of human resource development*

Performance describes the overall success rate of a person during a certain period in the duty compared to the wide range of possibilities, such as the standard of the work, the target or targets or criteria that have been determined in advance and have been agreed on (19). Furthermore, as proposed by Noe et al. (2008), the notion of performance includes the willingness of a person or group of people to do something to improve activities in accordance with their responsibilities with the expected results. The performance of specific functions does not stand alone, but in terms of job satisfaction and rates of return, is influenced by the skills, abilities, and individual traits. According to Simamora (2005), performance is essentially determined by three factors: (1) the ability (Luthans; Robbins and Mary 2002) factors that affect performance, which in this case are related to, among others, to human resource development, the environment, management, work stress, and motivation, leading to the following hypothesis:

H4: There is an influence of the K3 management system and work stress on performance management related to human capital development through motivation with K3 as an intermediate variable.

3 RESEARCH METHOD

This study used quantitative research methods. The results of the initial data analysis will be the basis of future research policy direction. We know that government policies are rooted in what's happening in a particular field of human resources development policy that is directly related to employees, so a company is sometimes less able to monitor anyone who has additional competence.

The results received from the scale of research subjects were then tabulated and we made a distribution of frequencies, which are grouped prior to the class interval, from lowest to highest with LPS.

4 RESULTS

Table 1 shows the results of descriptive data collected from Central Java and Yogyakarta area consisting of an area of Semarang, Purwokerto and Yogyakarta.

4.1 *Semarang area*

The number of descriptive respondents of the PLN Francisco area office was as many as 46 people, so that the class interval was 12.27 and the average score was 30.40. It shows that the average score of respondents' perceptions of the variable performance of human resources development in the Office of PLN Semarang at 30.40, which is within the Average category. This means that the variables of human resource development are still needed in order to enhance organizational performance.

The K3 management motivation variable indicates that the respondents' perception of the motivation variable K3 management in the Office of PLN Semarang has an average score of 33.70, which is within the Average category. This suggests that the K3 motivation management in the Office of PLN Semarang region is good enough.

Job stress variables showed that respondents' perceptions of the variable work stress in the Office of PLN Semarang gave an average score of 27.00, which is within the Average category. This shows that the employees in the Office of PLN Semarang region did not experience excessive stress in completing the tasks assigned to them.

K3 system variables showed that respondents' perceptions of variable K3 management system in the Office of PLN Semarang gave an average score of 35.27, which is within the High category. This shows that the management system of K3 at PLN Semarang Regional Office has been implemented very well.

4.2 *Purwokerto area*

The number of descriptive respondents of the PLN office Purwokerto area was 16 people. HR development performance variables show that the average score of respondents' perceptions of variable human resource development in the Office of Purwokerto amounted to 11.10 PLN, which is within the Medium category. This means that the PLN office employees' perception of human resources development of companies of the Purwokerto area so far has been quite good.

The K3 management motivation variable indicates that the respondents' perceptions of the motivation variable K3 management in the Office of PLN Purwokerto gave an average score of12.70, which is within the Medium category. This suggests that the motivation provided by K3 management in the Office of PLN Purwokerto area is good enough.

Job stress variables showed that respondents' perceptions of the variable work stress in the Office of PLN Purwokerto gave an average score of 10.20, which is within kategoriSedang.

Table 1.

Area	Number of subjects	Gender		Age				Level of education				
		L	P	≤30	31–40	41–50	≥50	SLTA	IN	DIII	S1	S2
Semarang	46	36	10	27	14	2	3	11	4	11	18	2
Purwoerto	16	14	2	13	2	0	1	7	3	4	2	0
Yogyakarta	18	18	0	10	5	1	2	9	6	1	2	0
Total	80	60	12	50	21	3	6	27	13	16	22	2

This shows that the employees in the Office of PLN Purwokerto region did not experience excessive stress in completing the tasks assigned to them.

K3 system variables indicate that respondents' perceptions of the variable K3 management system in the Office of PLN Purwokerto gave an average score of 12.02, which is within the High category. This shows that the management system of K3 in the Office of PLN Purwokerto area has been applied very well

4.3 Yogyakarta area

The number of descriptive respondents of the PLN office area of Yogyakarta was as many as 18 people. The variable of human resource development shows that the average score of respondents' perceptions of the variable human resource development in the Office of PLN Yogyakarta is 12.60, which is within the Average category. This means that employees of the Office of PLN Yogyakarta region perceive human resource development done by the organization or company so far has been quite good.

The K3 management motivation variable indicates that the respondents' perceptions of motivation variable K3 management in the Office of PLN Yogyakarta gave an average score of 13.23, which is within the High category. This suggests that the motivation provided by K3 management in the Office of PLN Yogyakarta region is good.

Job stress variables showed that respondents' perceptions of the variable work stress in the Office of PLN Yogyakarta gave an average score of 12.20, which is within the Average category. This shows that the employees in the Office of PLN Yogyakarta region did not experience excessive stress in completing the tasks assigned to them.

K3 system variables indicate that the respondents' perceptions of K3 management system variables in the Office of PLN Yogyakarta gave an average score of 14.12, which is within the High category. This shows that the management system of K3 in the Office of PLN Yogyakarta region has been implemented very well.

REFERENCES

Anderson, Erin, and Oliver, Richard L.. 1987. Perspectives on Behavior-Based versus Outcome-Based Salesforce Control Systems. *Journal of Marketing* 51 (October).

Djati, S. P. 2009. Variabel Anteseden Organizational Citizenship Behavior and Its Effect on Service Quality pada Perguruan Tinggi Swasta di Surabaya. *Jurnal Aplikasi Management*. 7(3).

Eisenberger, R., Armeli, S., Rexwinkel, B., Lynck, P. D., & Rhoades, L. 2001. Reciprocation of Perceive Organizational Support. *Journal of Applied Psychology*. 86(1).

Gallagher, C. 2001. *Health and Safety Management System: An Analysis of System Types and Effectiveness*. Upper Saddle River, NJ: Prentice-Hall.

George, J. M. 1991. State or Trait: Effects of Positive Mood on Prosocial Behaviors at Work. *Journal of Applied Psychology*.

Greer, Charles, R. 1995/ *Strategy and Human Resources, A General Managerial Perspective*. Upper Saddle River, NJ: Prentice-Hall.

Hasibuan, Malay S. P. 2009, *Human Resource Management*, Earth Literacy, Jakarta.

Jonathan, K., & Rosemary, W. M. 2016., Maintaining the Health and Safety at Workplace: Employee and Employer's Role in Ensuring a Safe Working Environment Journal of Education and Practice, www.iiste.org.

Ivancevic, J. M. 2001. *Human Resource Management*. New York: McGraw-Hill.

Noe, R., et al. 2008. *The Human Resource Management: Achieving Competitive Advantage, Book 1*. Interpretation. Jakarta: David Widjaja, Salemba Four.

Peraturan Perundang-Undangan Nomor 70 tahun 2012/ *Occupational Health and Safety*.

Robbins, S., & Mary, C. 2002. *The Management, Interpretation: Bob Sabran & Devri Barnadi Putera*. Jakarta: PT Gelora Literacy Primary.

Simamora, H. 2005. *Manajemen Sumber Daya Manusia*. Yogyakarta: STIEYKPN.

Sucipto, C. D. 2014. *The Occupational Safety and Health*. Yogyakarta: Goshen Publishing.

Werner, J. M., & DeSimone, R. L. 2009. *Human Resource Development*. Boston: South-western Cengage Learning.

Emerging Trends in Psychology, Law, Communication Studies, Culture, Religion,
and Literature in the Global Digital Revolution – Setiawan & Rahmawati (eds)
© 2020 Taylor & Francis Group, London, ISBN 978-1-03-224216-3

The influence of WhatsApp on improvements for fish farmers: A lesson from Semarang City, Indonesia

F. Apresia, T. Elfitasari & T. Susilowati
Faculty of Fisheries and Marine Science, Diponegoro University, Semarang, Indonesia

ABSTRACT: This study compares the use of WhatsApp by fish farmers, between users and non-users and the improvement of these fish farmers' financial conditions and aquaculture knowledge. Questionnaires were distributed to 60 fish farmers: 30 who already used WhatsApp and 30 who did not. Result shows that fish farmers who were WhatsApp users showed a higher score in both financial conditions and aquaculture knowledge than non-WhatsApp users. This indicates that fish farmers who use WhatsApp have higher knowledge and have better financial conditions than fish farmers who are not WhatsApp user.

1 INTRODUCTION

Mobile phones are no longer considered a luxury item. Many people across the world are using mobile phones, no matter their income, including small-scale fish farmers in cities and rural areas. As mobile phones can provide access to many software applications and social media, this makes them more attractive. WhatsApp is one of many popular social media applications currently offered on the market. Worldwide usage of WhatsApp was reported to have reached 1 billion users in 2016 (Kumar and Sharma, 2017). WhatsApp social media has made it easier for small-scale fish producers in Semarang City to interact with each other and find a quick solution to aquaculture problems. A small producer can also widen their business network and market using this application. WhatsApp users can download the application for free, and they can send message, pictures, photos, videos, audio messages and make video calls without charge if there is an internet connection. Furthermore, they can communicate with several people at once in a group chat (Rusni and Lubis, 2017).

1.1 *Fish farmers' aquaculture knowledge and financial conditions*

The two aspects explored in this research measured the aquaculture knowledge and financial conditions of fish farmers. The aquaculture knowledge aspect emphasizes their knowledge of aquaculture techniques, problem-solving abilities, and knowledge on how fish farmers become familiar with new information resources. The financial conditions aspect measured the prosperity of fish farmers, how they distribute a fish product or marketing system, and their social network.

1.2 *Fish farmers and WhatsApp*

Fish farmers using WhatsApp in Semarang in this study were 152 farmers distributed along 13 sub-districts, where there are only three extension services in each subdistrict (Semarang city Fisheries and Marine Office, press communication). The ratio of extension services to fish farmers is 4:1. However, each extension service is not only obliged to assist fish farmers, but are also required to support fishers and fish processors in the same sub-district. The limited number of extension services has made it difficult for the extension services to effectively help and reach each fish farmers location personally. This is one of the reasons that extensions services in Semarang city decided to create a WhatsApp group to make the communication with the fish farmers easier and faster. Intensive support by the extension services is needed in order to facilitate behavioral changes in the person that requires assistance (Safrida, Makmur and Fachri, 2015). This way, the function of an extension service is not only limited to transferring knowledge and information, as it can continuously develop good relationships and eventually create changes in other people.

Fish farmers that belong to the WhatsApp group will get information much faster, easier, and unlimited by time or distance. WhatsApp groups have been used by extension services in rural areas as media to share information, problems, and finding solution through discussions (Thakur, Chander and S. K. Sinha, 2017; Thakur, Chander and S. Sinha, 2017; Thakur, Chander and Katoch, 2018; Thakur and Chander, 2018).

2 METHODOLOGY

This research employed a quantitative method through questionnaire distribution to two groups of fish farmers: (1) fish farmers that use WhatsApp and (2) fish farmers that do not use WhatsApp. Variables used in this research were:

a. Independent variable (X1) fish farmers that use WhatsApp.
b. Independent variable (X2) fish farmers that do not use WhatsApp.
c. Dependent variable (A) knowledge enhancement of fish farmers
d. Dependent variable (B) financial conditions of fish farmers

Respondents were chosen using a purposive random sampling method to ensure a balance between the number of WhatsApp users and non-users. The population was also limited to fish farmers of freshwater, brackish water, and ornamental fish. However, age and education background were not limited. The data collected were then analyzed using SPSS software.

2.1 *Demographical condition of respondents – Age*

The ages of respondents are divided into five groups, as shown in Table 1

Table 1 shows that non-users of WhatsApp are dominated by one age group (41–50), whereas users of WhatsApp are more evenly distributed between age groups.

2.2 *Demographical condition of respondents – educational background*

The educational background of the respondents is shown in Table 2.

Table 2 shows that non-users of WhatsApp have a lower educational background than WhatsApp users. This indicates that lower-educated farmers use technology less often than farmers with higher levels of education.

Table 1. Respondents' age.

Age group	Non-users of WhatsApp	Users of WhatsApp
21 – 30	0 %	20 %
31 – 40	24 %	17 %
41 – 50	50 %	27 %
51 – 60	16 %	36 %
> 60	10 %	0 %
TOTAL	100%	100%

Table 2. Educational background of respondents.

Education	Non-users of WhatsApp	Users of WhatsApp
Primary	23.3 %	3 %
Junior High	33.3 %	7 %
Senior High	40 %	50 %
Higher Edu	3.3 %	40 %
TOTAL	100%	100%

Table 3. Respondents' aquaculture commodity.

Commodity	Non-users of WhatsApp	Users of WhatsApp
Freshwater	67 %	47 %
Ornamental	0 %	33 %
Brackish	33 %	20 %
TOTAL	100%	100%

2.3 *Demographical condition of respondents – commodity*

The aquaculture commodity of respondents was limited to freshwater, brackish water, and ornamental fish, as shown in Table 3.

Table 3 shows that WhatsApp users come from all three commodities (freshwater, brackish water, and ornamental). Interestingly, no non-users were farmers of ornamental fish. All farmers of ornamental fish are more advanced and have utilized WhatsApp for their business.

3 RESULTS AND DISCUSSION

3.1 *Aquaculture knowledge*

Aquaculture knowledge consists of three aspects: knowledge of aquaculture techniques, problem-solving abilities, and knowledge on how fish farmers become familiar with new information resources. The result of the data collection is provided in Table 4.

Table 4 shows that fish farmers who use WhatsApp have a higher overall score (30.8) in aquaculture knowledge compared to non-users (29.8). This indicates that fish farmers who are WhatsApp users have higher aquaculture knowledge compared to non-users. This result was as expected, as WhatsApp users can gain new information easier and faster. Even in an individual score of aquaculture techniques, problem-solving and new information, the score for WhatsApp users is consistently higher than non-users.

Table 4. Mean score of aquaculture knowledge of respondents.

	Mean score			
	A1 Aquaculture Technique	A2 Problem-solving	A3 New information	Total
Non-users of WhatsApp	18.4	5.5	5.3	29.2
Users of WhatsApp	19.3	5.8	5.7	30.8

Note: Total maximum score is 40 (A1: 24; A2: 8; A3: 8)

Table 5. Mean score of the financial condition of respondents.

	Mean score			
	B1 Prosperity	B2 Marketing	B3 Social network	Total
Non-users of WhatsApp	10.5	8.4	7.5	26.4
Users of WhatsApp	10.7	8.6	8.5	27.8

Note: Maximum score is 40 (B1: 16; B2: 12; B3: 12)

3.2 *Financial conditions*

The financial condition of fish farmers is based on three aspects: the prosperity of fish farmers, how they distribute a fish product or marketing system, and their social networks. The data collection result is shown in Table 5.

Similarly to the indicator of knowledge, the financial condition score of fish farmers that are WhatsApp users is higher (27.8) compared to non-users of WhatsApp (26.4). The higher score is also consistently found in all aspects of financial conditions: prosperity, marketing, and social networks. This result indicates that WhatsApp has a positive impact for fish farmers in improving their financial condition, including their welfare, access to a broader market, and social networking.

4 CONCLUSION

This research has confirmed that WhatsApp shows a positive impact on fish farmers, regarding their aquaculture knowledge and financial conditions. Both indicators show that the scores for fish farmers who are WhatsApp users are consistently higher than their non-user counterparts. The result of this research indicate that fish farmers who use WhatsApp retain higher levels of aquaculture knowledge and experience better financial condition than fish farmers who do not use WhatsApp. Therefore, it is advisable to extend the service to broadly utilize social media such as WhatsApp to assist fish farmers more effectively through the creation of WhatsApp groups.

REFERENCES

Kumar, N. and Sharma, S. 2017. Survey Analysis on the usage and Impact of Whatsapp Messenger. *Global Journal of Enterprise Information System*, 8, 3, 52–57.
Rusni, A. and Lubis, E. E. 2017. Penggunaan Media Online Whatsapp dalam Aktivitas Komunitas One Day One Juz (Odoj) dalam Meningkatkan Minat Tilawah Odojer di Kota Pekanbaru. *Jurnal Online Mahasiswa Fakultas Ilmu Sosial dan Ilmu Politik Universitas Riau*, 41, 1–15.
Safrida, S., Makmur, T. and Fachri, H. 2015. Peran Penyuluh Perikanan Dalam Pengembangan Sektor Perikanan Di Kabupaten Aceh Utara. *Jurnal Agrisep*, 16, 2, 17–27.
Thakur, D. and Chander, M. 2018. Effectiveness of whatsapp for sharing agricultural information among farmers of Himachal Pradesh. *Journal of Hill Agriculture*, 9, 1, 119–123.
Thakur, D., Chander, M. and Katoch, V. 2018. WhatsApp Model for Farmer Led Extension: Linking Actors and Generating Localized Information for Farmers. *Asian Journal of Agricultural Extension, Economics & Sociology*, 1–8.
Thakur, D., Chander, M. and Sinha, S. 2017. Whatsapp for farmers: Enhancing the scope and coverage of traditional agricultural extension. *International Journal of Science, Environment*, 6, 4, 2190–2201.
Thakur, D., Chander, M. and Sinha, S. K. 2017. A Scale to Measure Attitude of Farmers towards Social Media Use in Agricultural Extension. *Indian Research Journal of Extension Education*, 17, 3, 10–15.

*Emerging Trends in Psychology, Law, Communication Studies, Culture, Religion,
and Literature in the Global Digital Revolution – Setiawan & Rahmawati (eds)*
© 2020 Taylor & Francis Group, London, ISBN 978-1-03-224216-3

The pragmatic force of expressive speech acts of the Banyumasan Javanese language in sale and purchase transactions in Pasar Wage, Purwokerto Timur District, Banyumas Regency

M. Riyanton, Mustasyfa Thabib Kariadi & Etin Pujihastuti
Universitas Jenderal Soedirman, Purwokerto Utara, Indonesia

ABSTRACT: This research reviews expressive speech acts of rejection in the Javanese language in sale and purchase transactions in *Pasar* Wage. The problem studied in this research is the pragmatic force of expressive speech acts of rejection in Javanese language in sale and purchase transaction in *Pasar* Wage. The research objective was to describe the pragmatic force of expressive speech act of rejection in Javanese language in sale and purchase transactions in *Pasar* Wage, Purwokerto Timur District, Banyumas Regency. This qualitative descriptive research takes oral data in the form of speeches containing expressive speech acts of rejection in the Javanese language. The research data are collected from informants, in the form of speeches containing expressive speech acts of rejection. The research population is Javanese speeches containing expressive speech act of rejection. The advanced techniques employed are free-listening-participating-speaking (SBLC), listening-participating-speaking (SLC), recording and noting techniques. Contextual and equivalence methods are employed for analysis. The pragmatic forces resulting from the expressive speech acts of rejection in sale and purchase transactions in *Pasar* Wage are rejection in transaction and acceptance in transaction. The results of pragmatic force are influenced through the politeness principle and cooperation principle.

Keywords: speech act, pragmatic force

1 INTRODUCTION

Speech acts are very important in pragmatics, for stating something so that a speaker's intention is known by the listener.

There are eight types of speech act (Wijana, 1996): direct and indirect, literal and non-literal, literal direct, literal indirect, non-literal direct, and non-literal indirect. Therefore, determining the linguistic realization of which speech act is appropriate in certain situations requires knowledge of linguistic and strategic resources (Abolfathiasl, 2013). According to Searle (in Wijana, 1996) in a speech event there are three speech acts: locutionary, illocutionary and perlocutionary. A locutionary act is commonly known as the act of saying something. An illocutionary act serves to describe or inform about something, as well as to do something, commonly known as the act of doing something. A perlocutionary act is a speech act intended to influence the listener, known as the act of affecting someone.

Searle classifies illocutionary acts into assertive, directive, commissive, expressive and declarative (in Tarigan 2009), and categorizes speech acts into representative, directive, commissive, expressive, and declarative (in Rohmadi, 2004). Yule (2006) states that speech acts may be classified into declarative, representative, expressive, directive and commissive. Based on some experts' opinion and in view of the general function of speech acts, the speech acts may be classified as declarative, representative/assertive, expressive, directive and commissive. In addition, Searle's theory is then important for prostructuralist language to consider human factors in linguistic components in communication, situation and discourse. At the same time, this research is conducted through an external linguistic medium, based on a thorough understanding of the units and systems of relationship, or the significance of language inherent in a prior-system-structural scientific paradigm (Dementyev, 2016).

An expressive speech act is used to say something perceived by the speaker (Yule 2006). The verbs that mark expressive speech acts include, for example, those expressing gratitude, congratulations, condolence, and rejection, etc. Meanwhile, rejection means to not accept (from giving, yielding or granting), to deny, not justify, reduce, and drive out. Rejection is a speaker's expressive act towards something incompatible with their opinion or desire.

Rejection speech acts in Javanese often occur in trade transaction in traditional markets in Indonesia, such as in *Pasar* Wage. *Pasar* Wage, Purwokerto, Central Java has various forms of rejection in the bargaining process. Javanese culture inherent in this society's daily life highly affects such rejections.

Based on initial analysis on speech acts, it is interesting to study expressive speech acts of rejection in sale and purchase transactions, particularly in *Pasar* Wage in order to examine the pragmatic power.

2 METHOD

A research method is used to solve a problem. According to Kridalaksana (2008), the method is a way to approach, observe, analyze and explain a phenomenon. This article will describe the (a) type of research, (b) research location, (c) research data, (d) population and sample, (e) data collection method, (f) data analysis method, and (g) method used to present the data analysis result. This study used descriptive qualitative research. Descriptive research may be defined as a procedure to solve a problem by illustrating/depicting the condition of the research subject/object (people, institution, society, etc.) now based on observable facts or as is (Nawawi, 1991). Qualitative research is conducted according to scientific background or on the context of wholeness. Qualitative methodology is a research procedure which produces descriptive data in the form of people's written or oral words and observable behaviors, based on facts or phenomena actually occurring in the field. A pragmatic approach is employed in this type of research. Descriptive qualitative research is conducted to solve a problem investigated as a whole, by describing facts and phenomena in the field.

This research was conducted in *Pasar* Wage, Purwokerto Timur District, Banyumas Regency. *Pasar* Wage is a district-level Dutch legacy market, initially built to fulfill the residents' household needs and for selling the crops of those who live in the surrounding areas.

The population is the research object. Population, in general, is all individuals with certain linguistic aspects (Subroto, 1992). The research population is all Javanese speech containing expressive speech acts of rejection in the Javanese language in sale and purchase transactions in *Pasar* Wage.

3 RESULT

Pragmatic force is the effect arising in the listener's mind of the force of a speech act. If the speaker and listener understand the speech intention one to another, the pragmatic force of speech may be stated as successful, and vice versa. Whether or not pragmatic force is successful is influenced by the use of politeness and cooperation principles by the speaker and hearer. Table 1 describes the pragmatic force existing in expressive speech acts of rejection in sale and purchase transactions in *Pasar* Wage.

Table 1.   Acceptance of pragmatic force.

O1	*Monggo mbak salake mbak, dijajal ndisit ya olih.*
	'Please, Sis, salak, tester is available.'
O2	*Pinten buk?*
	'How much does it cost?'
O1	*Mangewu.*
	'Five thousand.'
O2	*Sekilo?*
	'One kg?'
O1	*La jeruk saiki pitungewu ra gedhe.*
	'It is seven thousand for orange, but for small ones.'
O2	*Sekawan ewu mboten angsal?*
	'What about four thousand?'
O1	*Tambahi mangatus maning ya.*
	'Please five hundred more.'
O1	*rep sekilo apa rong kilo?*
	'Please, one or two kg?'
O2	*Sekawan ewu buk.*
	'For four thousand.'
O1	*Nggeh. Rong kilo?*
	'Yes. Two kg?'
O2	*Sekilo tok.*
	'Only one kg.'

Context

A speech event occurs between (O1) as the seller and (O2) as the buyer. The tone of emotion is normal, but there is a little emphasis from the listener. The speech intention is that O2 wants to buy *salak*, but there is a price negotiation. There is no third person. The speech starts with O1 offering her *salak*, which is then continued by O2. The topic is the price of *salak*. The instrument used is *ngoko* style Javanese language. The type of speech is informal. The register of language is direct oral discourse. The speech employs *ngoko* style Javanese language.

In the speech event above, there is expressive speech act of rejection. The expressive speech act of rejection made by O2 uses a maxim of humility. This maxim is marked with *krama* Javanese language used by O2 in the transaction, even if O1 always responds using *ngoko*. The reason for this is that, by age, O2 is younger than O1.

Although some speech acts express rejection, the pragmatic force induced is that O2 accepts O1's offer to buy her *salak*. This positive response is closely related to the use of the maxim of humility by O2, which leads to O1's slightly smooth response, even if she uses *Ngoko* Javanese language. The smooth response of O1 eventually makes O2 accept the price, even if she initially rejects it multiple times.

In the speech event above, O1 maximizes the maxim of politeness, which is humility, as proven with O1's speeches that she always uses *krama* language. O2 does not use the maxim of politeness, as observable from her speech response that she always uses *ngoko* languge. However, both participants maximize the cooperation principle, so that the pragmatic force finally produced is O2's acceptance to buy orange.

Table 2.   XXX.

O1	*Salake murah, limangewu.*
	'This salak is cheap, five thousand.'
O2	*ora kurang? Patangewu wis bu.*
	'Cheaper please? How about four thousand?'
O1	*ya wis, pirang kilo?*
	'Okay, how many kg?'

Context

The speech event above occurs between O1 and O2. The tone of emotion is normal. The speech intention is that O1 offers her *salak*, and O2 negotiates the price. There is no third person. The order of speech is started by O1 as the seller and continued by O2 as the buyer. The topic is about the price of *salak*. The instrument used by both is *ngoko* Javanese language. The type of speech is informal. The speech scene is the fruit stall at *Pasar* Wage. The register of language is oral discourse. The language used is *ngoko* style Javanese language.

The speech event above does not use maxim of humility. However, because there is a maxim of agreement, the price is accepted with only one time-offering. The reason is that because there is personal closeness between O1 and O2 previously, as marked with the sentence *'ora kurang? Patangewu wis bu'* 'Cheaper please? How about four thousand'. The sentence implies that O1 is close with O2, as marked with O1's greeting to O2 using the word dhe, which means female relative. The standard price is not too different to the price offered by O1, from five thousand to four thousand, and this is a reasonable negotiation. With the combination of politeness principle, the pragmatic force produced is that O1 accept the price of *salak* and buys one kg of *salak*.

Explanation

In the speech event above, O1 maximizes the maxim of politeness, which is humility, as proven with O1's speech that she always uses *krama* language. O2 does not use maxim of politeness, as observable from O2's speech response that she always uses *ngoko* language. O1 maximize the cooperation principle, so that the pragmatic force eventually produced is that O2 accepts and buy orange.

Table 3.   XXX.

O1	*Mbakone pira buk?*
	'How much the tobacco cost, mom?'
O2	*patang ewu mbak.*
	'Four thousand, sis.'
O1	*Tigangewu buk. Olih ora?*
	'How about three thousand, mom?'
O2	*Ngapurane mba, nek kiye malah olih loro setengah ewu mbak.*
	'I'm sorry, sis, but this other one is for two thousand and five hundred.'
O1	*O, yawis.*
	'Oh, okay.'

Context

There is a speech event between O1 and O2. The tone of emotion is normal. The speech intention is that O1 wants to buy tobacco, but there is a negotiation process, implying a rejection. There is no third person. The speech order starts from O1 and is continued by O2. The topic is tobacco price. The instrument used is *krama* style Javanese language. The taste of speech is normal and informal. The speech scene is the tobacco stall in *Pasar* Wage. The speech uses register of oral discourse and *krama* Javanese language.

In the speech event above, O1 and O2 uses maxim of humility as marked with the *krama* Javanese language. In the speech event above, O1 negotiates the price for three thousand, one thousand different from the price offered. However, O1's negotiation is rejected by O2, and O2 explains about the quality of tobacco and recommends other tobacco. Finally, O1 accept the price offered by O2 for four thousand Rupiahs, as marked with the word *o yawis bu*.

In the course, the speech event above always uses *krama* Javanese language until the end of the transaction. O2 even expresses her sorrow to O1 when O1 negotiates the price, as marked with the sentence '*Ngaturaken lepat mbak, nek niki malah angsal kalih setengah mbak*' 'I'm sorry, sis, but this other one is for two thousand and five hundred.' This is a sign of use of high politeness principle in the speech event above. Although O2 is far older than O1, but as a seller, she keeps using maxim of humility and sympathy with each buyer. The pragmatic force of the speech event above is that O1 accepts the price and there is a sale and purchase event of one ounce of tobacco.

Explanation

In the speech event above, O1 and O2 maximize the politeness principle, which is humility, as observable because they always use *krama* style. Besides using the politeness principle, both speakers also maximize the cooperation principle, so that the pragmatic force eventually means O1 accepts and buys the tobacco.

Table 4.   Rejection of pragmatic force.

O1	*Niki pinten?*
	'How much does this cost?'
O2	*Pitulas.*
	'Seventeen.'
O1	*Pitulas?*
	'Seventeen?'
	Kalih pak, sedasa.
	'Two sir, ten.'
O2	*Iki anu mbak, awet bianget, nganggo serampat kok.*
	'This is long lasting, since it uses serampat.'
O1	*Kalih lo pak, sedasa.*
	'Two sir, ten.'
O2	*Telung puluh mbak.*
	'Thirty sis.'
O1	*Kalihdasa pak.*
	'twenty sir.'
O2	*Apik mbak iki.*
	'This is good sis.'
O1	*Penglaris lah.*
	'For good luck.'
	Nggeh, ngko nek ora percaya neng lor enek siji nggone bojoku.
	'Yes, if you don't believe it, there is on in the north, my wife's.'
O2	*Sedasa pak, mboten pareng ta?*
	"Ten sir, it's no?'

Context

The speech event above occurs between O1 and O2. The tone of emotion is informal, with a little light conversation. The speech intention is that O1 wants to buy an item and there is a negotiation and rejection of price. There is no O3. The order of speech starts from O1 as the buyer who asks about the price of sandals, answered by O2 as the seller, and there is negotiation. The topic is the negotiation of the price. The instrument used is *ngoko* style Javanese language O2 and *krama* style Javanese language by O1. The type of speech is relaxed. The speech scene is around 10.00 WIB at the shoe seller's kiosk in *Pasar* Wage. The register uses direct oral discourse.

In the speech event above, there is an expressive speech act of rejection made by O1 using a maxim of humility. This maxim is marked with *krama* Javanese language frequently used by O1 in the transaction, even if O2 uses *ngoko* language. The reason is that O1 is younger than O2.

Although O1 has used the maxim of humility, the pragmatic force produced is rejection of price. The reason may be unsuccessful use of the maxim of agreement, which causes a refusal to buy sandals. The use of the maxim is inappropriate, as observable from the too long negotiation, but O1 keeps the principle of negotiating with 'ten thousand', while O2 also maintains the price that the sandals will be sold for fifteen thousand Rupiahs. This results in disagreement between both parties. Besides no maxim of agreement, this failure of pragmatic force is also caused by non-use of the maxim of sympathy. This is observable because O1 keeps negotiating using smooth language in a friendly way, O2 only responds using short sentence, as in the speech below:

Table 5. XXX.

O1	*Kalih lo pak, sedasa.*
	'Two sir, ten.'
O2	*Telung puluh mbak.*
	'Thirty sis.'

In the speech, O1 negotiates with low tone and smooth sentences, but O2 only responds with short sentences, that the sandals are sold for thirty thousand Rupiahs.

The last response of the speech event is that O2 does not respond to O1, who keeps trying to negotiate the price for ten thousand Rupiahs, and O1 leaves the kiosk. The pragmatic force is disagreement between both parties, until finally O1 does not buy O2's sandals.

Explanation

In the speech event above, O1 maximizes the politeness principle, especially with the maxim of humility. O1 always uses *krama* style Javanese language in her speech. O2 minimizes the politeness principle, as observable from O2's speech in response to O1 that he always uses *ngoko* Javanese language. Both minimize the cooperation principle, so that the pragmatic force produced is rejection of transaction.

4   CONCLUSION

The pragmatic forces produced in the expressive speech acts of rejection in the sale and purchase transactions in *Pasar* Wage are rejection of transaction and acceptance of transaction. The results of this pragmatic force are influenced by the use of the politeness principle and cooperation principle by O1 and O2.

REFERENCES

Abolfathiasl, H. (2013) Pragmatic Strategies and Linguistic Structures in Making Suggestions: Towards Comprehensive Taxonomies. *International Journal of Applied Linguistics & English Literature*, 2, 6.

Dementyev, V. V. (2016) Speech Genres And Discourse: Genres Study In Discourse Analysis Paradigm, *Russian Journal of Linguistics*, 20, 4, 103–121.

Nawawi, H. (1991) Metode Penelitian Deskriptif. Gajah Mada University Press. Yogyakarta. Kridalak-sana, Harimurti, *Kamus Linguistik*, Jakarta: Gramedia.

Rohmadi, M. (2004) Pragmatik Teori dan Analisis, *Yogyakarta: Lingkar Media Jogja*.

Subroto, D. E. (1992) Pengantar Metode Penelitian Llinguistik Struktural, *Surakarta: Sebelas Maret University Press*.

Tarigan, H. G. (2009) Pengajaran Pragmatik, *Bandung: Angkasa Bandung*.

Wijana, I. D. (1996) Dasar-dasar Pragmatik, *Yogyakarta: Andi*.

Yule, G. (2006) Pragmatik (Edisi terjemahan oleh Indah Fajar Wahyuni dan Rombe Mustajab), *Yogyakarta: Pustaka Pelajar*.

*Emerging Trends in Psychology, Law, Communication Studies, Culture, Religion,
and Literature in the Global Digital Revolution – Setiawan & Rahmawati (eds)*
© 2020 Taylor & Francis Group, London, ISBN 978-1-03-224216-3

Reconstruction of regulations dealing with the period of the presidents' term to minimize the abuse of power

Sukimin
Universitas Semarang, Semarang, Indonesia

ABSTRACT: Provisions of the constitution of the Republic of Indonesia have determined the term of office of the president or head of state as five years. Afterwards he or she can be re-elected in the same position, but only once. but this is a reproach for the President and Vice President to be re-elected in the next election contest. There is a concern that state facilities attached to the office might be used in preparation for re-election. The approach method used in this study is normative juridical with descriptive analytical research specifications to review legal reconstruction related to the governance of the President and Vice President for a period so that there is no longer the term incumbent and opposition in order to achieve honest and fair general elections.

Keywords: reconstruction, governance, President and Vice President, power

1   INTRODUCTION

In discussing the period of administration of the offices of President and Vice President, of course we did not forget the previous administration, namely President Suekarno and President Soeharto. History noted that Soekarno tried to make himself President for life by issuing MPRS Decree Number III / MPRS / 1963. Concerning the appointment of the Great Leader of the Indonesian Revolution, Bung Karno became President of the Republic of Indonesia for life, while Soeharto politicized Article 7 of the Constitution of the Republic of Indonesia in order to be re-elected without limiting the number of periods. Article 7 before the amendment follows: "The President and Vice President hold their positions for five years and thereafter re-elected." The article is clear that the positions of President and Vice-President can be appointed again after five years without any limitation on the number of times during the reign, so that post-reformation changes in provisions related to the government President and Vice President in Indonesia (Ketetapan Majelis Permusyawaratan Rakyat Republik Indonesia Nomor XIII/MPR/1998, 1998).

The provisions of the constitution of the Republic of Indonesia regulate the term of office of a head of state, namely the President and Vice President. In carrying out their duties, the President and Vice President hold positions for five years, after which they can be re-elected in the same term, only for one term. Reinforced in Article 169 letter n of Law Number 7 of 2017 concerning General Elections, the article follows: "Requirements to become a candidate for President and candidate for Vice President are: n. has never served as President or Vice President for 2 (two) times in the same position."

There are three concepts of limitation of governance, namely:

1. There is no second term (no re-election);
2. There must be no continuing term of office (no immediate re-election); and
3. Maximum term of office (only one re-election).

Actually there is fourth concept of the term of office, that is, there is no limitation on term of office (no limitation to re-election), but of course the last concept is not in accordance with a presidential system which certainly requires the limitation of presidential term.

The reasons for limiting the period of the President and Vice President can include the emergence of authoritarian attitudes and dictators, to a President vulnerable to abuse of power, to congestion of national governance regeneration, and to the possibility of individual cults.

2 PROBLEM

Based on the above description, those things are the background of the author to conduct an assessment with the formulation of the problem: Why is the Need for Reconstruction of the President's Government Regulations Regarding the Abuse of Power?

3 RESEARCH METHOD

Research activities are activities carried out by efforts to understand and solve problems scientifically, systematically, and logically. Methodology comes from the word methods, which means ways or paths, and logos, which means science. In connection with scientific endeavors, the method involves work problems in order to understand objects in the science that is concerned. The method used in this study is a normative juridical approach and the specifications used in this study are descriptive analytical because this study is expected to be able to obtain a clear, detailed, and systematic picture, whereas it is said to be analytical because the data obtained will be analyzed for solving problems according to applicable legal provisions. The purpose of the study uses analytical descriptive specifications to provide an overview of reality on objectively examined objects.

4 RESULT AND DISCUSSION

4.1 *Reconstruction of regulations for the President's administration regarding abuse of power*

The branch of executive power is the highest authority of the state administrative authority, or better known as the government system. In the State Law there are three systems of government, namely Presidential, Parliamentary, and mixed. The Indonesian state itself adheres to the presidential system where the position of head of state is not separate from the position of head of government: the head of state is not responsible to the parliament, but directly to the people who elected him or her.

The presidential system of government does not distinguish whether the President is the head of state or head of government, but only the President and Vice President with all their rights and obligations or their respective duties and authorities. The government system is of course very much related to the election of head of state or head of government. Indonesia has several times experienced the election of the head of state, in this case the President. The General Election is the implementation of a clean direct democratic order, so that the people can take a role to determine their leaders. The General Election itself is held every five years for the election of the DPR, DPD, Provincial DPRD, Regency/City DPRD, and the President and Vice President.

Regarding the administration of the President and Vice President, the provisions of the State Constitution of the Republic of Indonesia regulate the term of office of a head of state, namely the President and Vice President, in carrying out their duties. Article 7 of the 1945 Constitution states "The President and Vice President hold positions for five years, after which they can be re-elected in the same position, only for one term of office," but this article is considered open to multiple interpretations, including whether the two periods are consecutive, so that they can re-nominate themselves as President and Vice President after having stopped for one period. The provisions of Article 7 of the Constitution have been subjected to judicial review at the Constitutional Court, but in the Number decision 40 / PUU-XVI / 2018 requests for applicants cannot be accepted.

It is only natural that a leader of the head of state wants to take office again after his term is over, and this can lead to an authoritarian leader. The authoritarian political system that was built was always included in the reproaches of the 1945 Constitution, so there was a need to build a democratic political system by amending the 1945 Constitution, or in other words, a need for a legal reconstruction related to the administration of the President and Vice President. In line with that, Article 7 of the 1945 Constitution has a defect: the President and Vice President, of course, will have a great opportunity to be re-elected in the next election contest (Undang-Undang Dasar Negara Republik Indonesia Tahun 1945, 1945).

The history of democracy notes that in 2009, the incumbent president Susilo Bambang Yudhoyono was able to maintain the election contestation, and we compare it with the 2019 election where President Joko Widodo can also maintain general elections. From the KPU statistical data, the acquisition of a number of votes for the election of the President and Vice President of the 2009 election president broke down Susilo Bambang Yudoyono in pairs with Boediono to win a landslide victory with 60.80%, while the 2019 president broke out Joko Widodo paired with Ma'aruf Amin to win the election by getting 55.50%.

Opportunities for President and Deputy President Batalana are greater to win the general election. Of course, we can see this more mature preparation of a president and political parties in campaigning themselves long before the election stage starts, to achieve their goals in political ways. It is desirable to have a situation wherein there is not the possibility that even the heads of state, in this case the President and Vice President, can abuse the power and use state facilities attached to the position prepare for re-election.

Reporting and publications to the public about their successful governance while serving as President and Vice President, certainly can affect the voters who later must be able to choose an incumbent partner, so the chances of escaping the opposition will be smaller and require a very hard struggle. The is a need to conduct legal reconstruction related to the administration of the Presidental and Vice Presidental period so that there is no inconsistency and opposition so that general elections that are honest and fair can be achieved.

Another factor is the impact on the performance of the President and Vice President, where in the during first to fourth years of the term there is still enthusiasm to work, but toward the end of the government period, of course the President and Vice President will prepare themselves to contest the election. Researchers argue that a term of office for President and/or Vice President that is only for one period will be able to realize fair elections and minimize the occurrence of misuse of office, related to legal reconstruction, then Article 169 of the Law Number 7 of 2017 concerning General Elections (Undang-Undang Nomor 7, 2017), and amend amendments to Article 7 of the 1945 Constitution of the Republic of Indonesia.

5 CONCLUSION

Provisions related to the administration of the President and Vice President are regulated in Article 7 of the Basic Law which states "The President and Vice President hold positions for five years, and after that they can be re-elected in the same position, only for one time." The President and Vice President will certainly have a great opportunity to be re-elected in the next election contest, so the President and Deputy President may be able to misuse power and state facilities attached to the office to influence re-election. This makes it necessary to persue legal reconstruction related to the administration of the President and Vice President for one period so that there are no longer the terms incumbent and opposition and general elections that are honest and fair can be achieved.

REFERENCES

Asshiddiqie, Jimli. Pengantar Ilmu Hukum Tata Negara. Jakarta: Raja Grafindo Persada. 2015.

Indarj.perkembanganPemilihanPresidendanWakilPresiden di Indonesia, Jurnal, Masalah-Masalah Hukum: Universitas Diponegoro, Jilid 47, No.1, Januari 2018.

Ketetapan Majelis Permusyawaratan Rakyat Republik Indonesia Nomor XIII/MPR/1998 (1998) *Pembatasan Masa Jabatan Presiden dan Wakil Presiden Republik Indonesia.*

Koentjaraningrat. Metode-Metode Penelitian Masyarakat. Jakarta: PT. Gramedia, 1998.

Mahfud, Moh MD. Politik Hukum di Indonesia. Jakarta: Raja Grafindo Persada. 2009.

Putusan Mahkamah Konstitusi Republik IndonesiaNomor 40/PUU/XVI/2018.

Undang-Undang Dasar Negara Republik Indonesia Tahun 1945 (1945) *Undang-Undang Dasar Negara Republik Indonesia Tahun 1945.*

Undang-Undang Nomor 7 (2017) *Pemilihan Umum.*

Zainuddin, Muhammad. PemahamanMetodePenelitianHukum (Pengertian, Paradigma, dan Susunan Pembentukan. Yogyakarta: CV. Istana Agency. 2019.

Emerging Trends in Psychology, Law, Communication Studies, Culture, Religion, and Literature in the Global Digital Revolution – Setiawan & Rahmawati (eds)
© 2020 Taylor & Francis Group, London, ISBN 978-1-03-224216-3

Sociocultural theory applied to collaborative learning in writing strategies

Testiana Deni Wijayatiningsih
Universitas Muhammadiyah Semarang, Semarang, Indonesia

Mustasyfa Thabib Kariadi
Universitas Jenderal Soedirman, Purwokerto Utara, Indonesia

ABSTRACT: Teaching writing is a crucial skill in promoting learners' English proficiency. Writing needs a sequence of stages, done step by step. These activities were used with English Department students. Although they followed some sequences of writing stages, they had low proficiency in writing paragraphs. In order to acquire productive skills in writing, this study undertook qualitative research on applying sociocultural theory to collaborative learning in writing strategies. The research objective was to describe implementing sociocultural theory to collaborative learning in writing strategies. This research used a qualitative method that describes the application of sociocultural theory to collaborative learning in writing strategies. This study also used observation to reveal the application of sociocultural theory to collaborative learning in writing strategies. The result shows that sociocultural theory has made a great impact on the learning and teaching profession as it is included in the process of teaching and learning writing, and this skill is regarded as social process. Further, the perspective of sociocultural becomes the foundation in providing scaffolding in teaching writing. Likewise, this learning process also involves collaboration in writing lessons, as well as lecturers' reflections on their teaching and learning activities in the classroom, as much as the teaching and learning writing as social process.

Keywords: sociocultural theory, collaborative learning, writing strategies

1 INTRODUCTION

Teaching writing is a crucial skill in promoting learners' English proficiency. Writing needs a sequence of stages, done step by step: being motivated to write, getting ideas together, planning and outlining, making notes, making a first draft, revising, re-planning, redrafting, editing, and getting ready for publication (Hedge, 2015). These activities were used with English Department students. Although they followed some sequences of writing stages, they had low proficiency in writing paragraphs. In order to acquire productive skills in writing, this study undertook qualitative research on applying sociocultural theory to collaborative learning in writing strategies. The research objective was to describe implementing sociocultural theory to collaborative learning in writing strategies.

As stated by Vygotsky (1896–1934), there is a strong relationship between culturally organized experiences and learning. Therefore, Vygotsky focuses on language as a social and communicative activity. He also argues that higher level skills are the result of the child's learning of social-functional relationships, and then, in becoming literate, children learn the structures and processes inherent in socially meaningful literacy activities. In this way, processes that are initially mediated socially become resources available to the individual language user.

Those ideas are very useful in teaching and learning in general, as they emphasize how interactions between people become the most important mechanism by which learning and development occur. The key assumption is that the intellectual skills acquired by children are directly related to their interaction with adults and peers in specific problem-solving environments. In this sense, the interactions between people and individual psychological processes rely on an explicit and direct connection. Therefore, the role of the teacher, peers, and others in the teaching and learning process is pivotal, as others' development can be driven through their assistance in social bonds (Samana, 2013).

Further, Vygotsky also discusses a theory in teaching and learning writing, called the zone of proximal development (ZPD) (Nurfaidah, 2017). This theory emphasizes the influence of macro-structures and socio-political contexts as simultaneous constraints in human activity. Under this theory, teachers and learners' behavior can be analyzed and observed (Storch, 2018). Moreover, this is useful in identifying the teachers' practices for feedback on student work. The teachers' underlying motives and drive in giving certain feedback can be revealed. Similarly, students' ways of responding to feedback could also be investigated (Storch, 2018). It can be concluded that ZPD refers to the distance between the actual developmental level, as determined by individual problem-solving, and the level of potential development as determined through problem-solving under adult guidance or in collaboration with more capable peers.

Regarding teaching and learning writing, the activity theory operates on learners' participation in such an activity system, which requires the learners to use strategies when dealing with writing tasks (Zohreh, 2014). As feedback from the teachers operates on scaffolding schema, learners can locate strategies to develop more in their learning process, such as self-revision, asking for peer revision or other ways that foster the learners' independence and editing skills. Teachers, in turn, have the responsibility to provide feedback to the learners appropriate to their individual needs.

Meanwhile, in this language learning, Bruner (1966) in Judith (2007) views language as providing the basis for concept formation, as a tool for cognitive growth. Further, he sees writing as a powerful tool essential for thinking, and schooling as promoting the growth of reasoning abilities through training in the mastery of the written language. Bruner also believes that written language is particularly important in encouraging cognitive growth because it is abstract, and the referent is not as frequently present as it is during many forms of oral discourse. Meanwhile, the language of school is particularly important in developing abstract literacy skills, requiring students to go beyond the information given and to deal with possibilities and abstractions. Therefore, both Vygotsky and Bruner see language learning as growing out of a communicative relationship where the adult helps the child to understand as well as to complete new tasks. This means that those concepts are also applicable to the situation of teachers and learners because encouraging these types of thinking and reasoning can support learners' higher levels of cognitive development.

Raimes (1985) in Irwansyah (2016) states that writing is an effective way for learners to generate words, sentences, and chunks of discourse and to communicate them in a new language. Furthermore, Langan (2001) in Irwansyah (2016) describes writing as a process of discovery involving a series of steps, and those steps are very often a w journey. It can be concluded that writers do not discover what they want to write about until they explore their thoughts in writing.

2   METHOD

This research concerns a qualitative method that describes the application of sociocultural theory to collaborative learning in writing strategies. Qualitative research can also be described as an effective model that occurs in a natural setting and enables the researcher to develop a level of detail from being highly involved in the actual experiences (Creswell, 2003). This study used observation to reveal the application of sociocultural theory to collaborative learning in writing strategies.

3 FINDINGS AND DISCUSSION

To promote better writing activity in the classroom, teachers are required to choose approaches that accommodate time, students' needs, and practice. Thus, in Vygotsky's ZPD the more competent peers are able to support learners' development. In the case of learning writing under collaborative work, social interaction strengthens the learners' levels of shared meaning elaboration (Mayordomo & Onrubia, 2015). When collaborative activities take place in the ZPD, the more capable peer is encouraged to offer help. The process of negotiating, constructing and reconstructing knowledge and new meaning takes place within a natural shared knowledge construction phase. This means that teachers should be more concerned about their students when they are involved in group work, in such a way that the coordination of teamwork could enhance the process of collaborative knowledge construction and collaborative learning in writing strategies.

In writing strategies, Badger (2000) in Hyland (2003) further informs that the application of teaching writing has made use of multiple approaches or a process-genre approach (PGA). In other words, teachers have incorporated several approaches to teaching writing to help students. Practically, PGA incorporates the four teaching steps of genre-based approach, as follows.

3.1 Building Knowledge of the Field (BKOF)

All activities aimed at defining situations that will be used as the topic and placing it within a particular genre are also implemented through a brainstorming stage in the process-based approach. Furthermore, this stage prepares the students to anticipate the structural features of the genre from variation of relevant texts (Yan, 2005). Students need to know what the topic under discussion is in order to determine the specific topic they want to write about (Emilia, 2008).

Therefore, in this stage the lecturer does activities with the students, such as building cultural contexts, sharing experiences, discussing vocabulary, and grammatical patterns, etc. Moreover, the lecturer gives the students direct instructions because they are still progressively assimilating the task demands with the procedures for constructing the texts effectively. The lecturer tries to link the students' own experiences with the text type they are studying. In other words, the students are expected to bring their own experiences to the learning process. In the classroom practice, students have different experiences of a certain text type that can be shared.

3.2 Modeling of Text (MoT)

The modeling stage is meant to give students in-depth information about the text type they are learning, through the "stages of the genre and its key grammatical and rhetorical features" (Hyland, 2007). The provision of varied text sources of the genre for students are aimed at getting them to understand how the organization of the text (schematic structure) is developed to accomplish the purpose (Yan, 2005) and also the linguistic features of the genre.

Therefore, after completing the BKOF stage, the lecturer does MoT which is the second stage in GBA. In this stage, the lecturer and students discuss and explore a text and its key grammatical features. The main purpose is to ensure the students focus on the features of the target genre. Some examples of texts of a certain genre are analyzed, compared, and manipulated so the students understand the generic structure needed to produce good pieces of writing. For example, if the main purpose is students are able to produce procedural texts, then the short functional texts, conversations, and the monologues are developed with one main communicative purpose: giving instruction or direction. At this stage, there is an analysis and discussion of how and why a certain type of text is organized to express meaning. In conclusion, through text deconstruction, it is possible for students to analyze the text components.

3.3 Joint Construction of Text (JCoT)

According to Yan (2005), the goal of JCoT is to produce a final draft that provides a model for students to refer to when they work on their individual compositions in the independent construction stage. In this stage, the lecturer implements scaffolding techniques, giving the students initial explicit knowledge and guided practices. Here, the lecturer and the students work together, with the lecturer as a guide, to develop targeted texts until the students can do it by themselves. The students attempt to develop texts of the target genre with the help from the lecturer, who reduces their contribution gradually as learners gain greater control over their writing. In other words, the lecturer acts only as a facilitator for the shared writing activities and as a responder to the student writing. The main purpose is to make the students show their writing ability and confidence to write.

3.4 Independent Construction of Text (ICoT)

Students write individually through guidance provided by the teachers. Lecturer can decide the topic or students can choose freely a topic that is still relevant to the genre. Similar to the genre-based approach, the teachers' control decreases as students start to apply what they have learned (Hyland, 2007), but the teacher is available to help, clarify, or consult the process of writing. In this stage, after collaborating with peers and the teacher, the students are expected to be able to produce their own texts independently and smoothly. This final stage allows the students great opportunities to show what they have learned and to write a text independently while the lecturer observes and gives advice from the sidelines. The following activities are the done by the teacher in this stage: (1) establishing groups with suitable number of students to have a discussion; (2) giving each group a chance to develop ways of presenting or teaching their texts to other learners; (3) providing an opportunity for the students to summarize or make an overview of the lessons of the day; (4) motivating the students to re-learn the teaching materials and/or complete home tasks independently or in groups; (5) giving each group a chance to present their lessons and receive appreciation for their efforts.

In closing the activity of learning, the lecturer does the following: (1) asks the students to summarize what they have learned either in individually or in groups; (2) asks the students' opinions or feelings on what they had just learned; (3) discuss the teaching materials and/or tasks for the next lesson.

4 CONCLUSION

Sociocultural theory has made a great impact on the learning and teaching profession as it is included into the process of teaching and learning writing in which this skill is regarded as a social process. Further, the sociocultural perspective becomes the foundation of scaffolding in teaching writing. Likewise, this learning process also involves collaborative work activity in writing lesson,s as well as lecturers' reflections on their teaching and learning activities in the classroom, as much as the teaching and learning writing as social process.

REFERENCES

Creswell, J. (2003) *Research design: Qualitative, quantitative and mixed methods approaches* (2nd ed.). Thousand Oaks, CA: SAGE Publications.
Emilia, E. (2008) Pendekatan genre-baseddalam Kurikulum Bahasa Inggristahun 2006: Penelitian tindakan kelasdi sebuah SMP Negeri di Bandung. *Bandung: Jurusan Pendidika Bahasa Inggris FPBS UPI*.
Hedge, T. (2005) *Writing*. Oxford: Oxford University Press.
Hyland, K. (2003) *Second Language Writing*. New York: Cambridge University Press.
Hyland, K. (2007) *Genre and Second Language Writing*. Michigan: The Unversity of Michigan Press.
Irwansyah. (2016) Genre Based Approach: A Way to Enhance Students Writing Ability. *English Education: Jurnal Tadris Bahasa Inggris*, 9, 1, 74–88.

Kim, Y., & Kim, J. (2005) Teaching Korean university writing class: Balancing the process and the genre approach. *Asian EFL Journal*, 1, 1, 68–89.

Mayordomo, R. M., & Onrubia, J. (2015) Work coordination and collaborative knowledge construction in a small group collaborative virtual task. *The Internet and Higher Education*, 25, 96–104.

Nurfaidah, S. (2017) Vygotskys Legacy on Teaching and Learning Writing as Social Process, *LANGKAWI Journal*, 4, 2, 149–156.

Raimes, A. (1985) What unskilled ESL students do as they write: a classroom study of composing. *TESOL Quarterly*, 19, 2, 229–258.

Samana, W. (2013) Teachers and students scaffolding in an EFL classroom. *Academic Journal of Interdisciplinary Studies*, 2 8, 338–343.

Storch, N. (2018) Written corrective feedback from sociocultural theoretical perspectives: A research agenda. *Language Teaching*, 51, 2, 262–277.

Yan, G. A. (2005) Process-genre model for teaching writing. *English Teaching Forum*, 43, 3, 18–26.

Zohreh. (2014) ZPD, Tutor; Peer Scaffolding: Socio Cultural Theory in Writing Strategies Application. *Procedia - Social and Behavioral Sciences*, 1, 1, 1771–1776.

Author Index

Abdurrohman, M.F 83
Anggraheni, D 41
Apresia, F. 107
Astanti, D.I 45

Bastos, J.C 73
Basyir, J. 60
Bocar, A.C. 14

de Yong, S. 49

Elfitasari, T. 107

Hakim, S.N 5
Hanifa, M. 5
Heryanti, B.R. 18, 45

Irawan, M.N. 78

Janie, D.N.A. 73
Juita, S.R. 45

Kadiyono, A.L. 83

Kariadi, M.T. 111, 121
Kridasaksana, D. 22
Kurniawan, Y. 102

Lestari, A.P. 90

Marta, R.F. 60
Mulyani, T. 56, 68, 95
Muryati, D.T. 56, 68, 95

Nugraha, A.K.N.A. 9

Priyanto, S.H. 9
Pujiastuti, E. 18, 56, 68, 95
Pujihastuti, E. 111
Purwaningtyastuti 26
Purwasih, J.H.G. 98
Putri, Rr C.W. 64

Rachmawati, A.W. 14
Rahmawati, S. 14
Rego, B.C. 73
Riyanton, M. 111

Sari, S.M. 49
Satrio, D. 9
Savitri, A.D. 26
Setiawan, Y.B. 60, 90
Sihotang, A.P. 29
Solikhati, S. 90
Sukimin 117
Sulistiobudi, R.A. 83
Susilowati, T. 107

Triasih, D. 18

Wahyuni, H.I. 36
Wai, T.K. 78
Widhiastuti, H. 1, 102
Widianto, A.A. 98
Wiendijarti, I. 36
Wijayatiningsih, T.D. 121
Witjaksono, R. 36

Yuliasih, G. 102
Sukimin 117